CIRCUITS OF THE MIND

CIRCUITS
OF THE MIND

LESLIE G. VALIANT

New York   Oxford
OXFORD UNIVERSITY PRESS
1994

Oxford University Press

Oxford New York Toronto
Delhi Bombay Calcutta Madras Karachi
Kuala Lumpur Singapore Hong Kong Tokyo
Nairobi Dar es Salaam Cape Town
Melbourne Aukland Madrid

and associated companies in

Berlin Ibadan

Published by Oxford University Press, Inc.
200 Madison Avenue, New York, New York, 10016

Oxford is a registered trademark of Oxford University Press

Library of Congress Cataloging-in-Publication Data
Valiant, Leslie.
    Circuits of the mind / Leslie G. Valiant.
      p.   cm.
    Includes bibliographical references and index
    ISBN 0-19-508926-X
    1. Computational neuroscience
    2. Neural networks (Neurobiology)
    I. title QP383.3.V35 1994   612.8'2–dc20   94-20869
                                                          CIP

9 8 7 6 5 4 3 2 1

Printed in the United States of America
on acid-free paper

To Gayle, to Paul, and to Gregory

Preface

It has been said that science can be defined as any discipline in which a fool of this generation can go beyond the point reached by a genius of the last. It is perhaps surprising that any field of endeavor can satisfy this strict criterion, but there is little doubt that some do. In fields that do, it appears that continuous progress is possible for long periods almost routinely. In the established physical sciences, for example, a rich intellectual structure has been uncovered that reveals at any time a wide range of unsolved problems or puzzles. Solutions to these provide increased understanding of the field and further enrich the structure. As long as successful problem solving continues, progress is close to being guaranteed. The possibility of almost routine progress of this nature appears to be a fundamental aspect of science. Even if it is not the most celebrated aspect, it may be the most characteristic one.

The study of cognition and the mind has probably not yet matured to the stage where routine progress is possible or the given criterion of science satisfied. The questions that one would think of as central have yet to be found formulations that reduce them to problem solving. Although several sciences are encroaching on this area, it is questionable whether they have succeeded in occupying its center. The main task for the present, therefore, may be viewed as a prescientific one. What is the most promising way to proceed in order to find the intellectual structure within which at least some central questions can be formulated and reduced to problem solving?

This volume suggests one avenue. It places at the center of the investigation some simple tasks of memory and learning, and advocates that these tasks be investigated by means of detailed computational models. The primary content is technical. It consists of a definition of a model together with descriptions of how a variety of tasks can be realized by programming this model appropriately. For the purpose of motivating the choices made and evaluating what is demonstrated, additional material is included that references findings in cognitive psychology, neuroscience, and computer science. We have kept these references brief, and necessarily incomplete, so as not to divert attention from the primary theme. It is clear, however, that the ultimate value of the approach will depend on how successfully some of these connections can be made in the future.

The model that will be defined is that of a system of model neurons that will be called neuroids. Its ultimate purpose is to bring within the scope of analysis some significant aspects of cognition. There are several disciplines that have similar aims and it is pertinent to ask at the start how the neuroidal model relates to models currently used in these various fields. A very rudimentary answer that conveys at least some of the basic intentions of the model and the positioning of this study would go as follows. The models that neuroidal systems offer are intended to be more unified and computationally detailed than those typically used in cognitive psychology, more programmable and of broader functionality than those in the field of neural networks, at a higher systems level than in neuroscience, and subject to more realistic resource constraints than those typical in artificial intelligence. Thus in these various respects the model is intended to offer something to each of these areas.

Since the use of unambiguous models is advocated, formal notation is frequently used in the text. Formalisms are needed both for the purpose of expressing programs as well as for analyzing them. The neuroidal model may be viewed as a programming language and those familiar with standard programming languages will be able to devise their own programs for this model with little difficulty. For the purpose of analyzing the algorithms some elementary probability theory is used. This is done sparingly except in the one chapter toward the end. The essential points of each chapter are intended to be accessible to those motivated readers who do not wish to delve too deeply into either kind of formal notation. The mathematically sophisticated will know how to skip formulae. This skill is easy to practice for others also.

This project has taken a long time to come to fruition. A sabbatical leave at Oxford University allowed me to write a preliminary paper. Subsequent work on this book was completed at Harvard University and at the NEC Research Institute in Princeton, New Jersey. I am grateful to all three institutions for providing me with the opportunity for pursuing such a speculative line of research. I am also grateful to the National Science Foundation, the Office of Naval Research, and the Advanced Research Project Agency for funding my research at Harvard in this and other areas over many years.

In the course of these years I have had valuable conversations with numerous colleagues on the range of topics that this volume addresses. My gratitude goes to those who provided an informative comment, a word of encouragement, or, in several cases, a concerned look. A few devoted many hours to reading all or parts of a preliminary version of this manuscript and provided some valuable comments and advice. I am particularly indebted in this regard to Dana Angluin, Eric Baum, Richard Herrnstein, Philip Johnson-Laird, Roni Khardon, Christof Koch, Hector Levesque, Wolfgang Maass, Dan Roth, and David Waltz.

The manuscript in its many versions was typeset initially in LATEX by Carol Harlow. I am grateful for her endless patience and good humor.

Finally, I have to thank my immediate family — Gayle, Paul, and Gregory — for providing inspiration in their various ways.

Contents

CIRCUITS OF THE MIND

Chapter 1
The Approach

Consider what happens when a person is exposed to a surprising juxtaposition of words as occurs sometimes in the title of a new book. The experience causes some adjustments in the brain, one must presume, since the presentation of the same sequence of words at a later time may elicit some form of recognition. Memorization of this kind requires few preconditions. The phrase to be memorized may be almost any combination of intelligible words, and the presentation of it may be very brief in time. Also, memorization has a certain robustness in the sense that it does not appear to interfere with unrelated knowledge previously memorized.

Even for an apparently simple phenomenon as this no widely accepted theory exists that can explain how the brain accomplishes it. Worse still, no theories appear to exist that explain how *any* mechanism that even remotely resembles the brain in structure and quantitative parameters could give rise to the variety of such basic phenomena that human brains appear to exhibit.

Our purpose in this monograph is twofold. The first is to show that this challenge can be formulated in a concrete manner. It is possible to attempt plausible specifications both of the cognitive functions that are to be explained, as well as of some computational models that capture the brain's basic capabilities. Such specifications help in contributing provocative questions about the central problems and make the difficulties harder to evade. We shall suggest some candidates for these cognitive functions and computational models. If these choices turn out to be right, then an algorithmic explanation will exist of how these functions are

1

indeed supported in the brain. If the choices are inaccurate then alternative choices of a similarly unambiguous nature will need to be found and explored.

The second of our purposes is to point out that the insights gained in recent decades from thinking about computation may provide a powerful new methodology for approaching the understanding of cognition and of its basis in the brain. Our thesis is that if the computational analogy has borne only limited fruit to date it is not because it is fundamentally flawed, but because it has not been taken far enough. By this we do not imply that the brain is at all similar to an electronic computer. What we mean is that the same methodology should be used for understanding the computational power of the brain as has been used for understanding the power of the wide variety of computational models that have been studied.

A computational account of the workings of the brain would consist of three parts. First, one needs a specification of the cognitive *functions* or tasks for which an explanation is sought. These could include rote learning or memorization, as illustrated in the above example, as well as recall, inductive learning, and any other task that is regarded as basic to cognition. The specifications need to be unambiguous. They also need to accurately capture essential aspects of the respective phenomena. Second, there has to be a description of the basic *model of computation*. In the cerebral context this would include definitions of the individual components that correspond to neurons, as well as of the connections through which they communicate with each other. Such a model would need to incorporate parallelism (i.e., simultaneous computational processing in many places) and hence differ from the current standard for electronic computation, the von Neumann model. Lastly, one has to specify the computational mechanisms or *algorithms* that enable the underlying model to realize the claimed functions or tasks.

Putting together such an account for even the superficially simplest tasks, such as memorization, becomes problematic as soon as we try to keep faith with the gross quantitative parameters that the brain is known to have. One major constraint is speed. Individual neurons in the cortex are slow. They are thought to have basic switching times of between 1 and 10 milliseconds. In contrast, humans can perform significant tasks of scene recognition in 100-200 milliseconds, tasks that no one knows how to perform by computer, even in principle, in any amount of time. In other words, in a few hundred steps at most the brain can do tasks that are currently beyond our imagination to even specify. The sparseness of the interconnections relative to the total number of neurons is another important constraint, since this limits the means by which neurons can communicate with each other. A third constraint is the finiteness of the number of neurons. Any proposal has to respect the actual bounds on this number.

The mystery is added a final twist by the issue of *learning*. Many recognition tasks that are easy for children but currently impossible for computers, such as identifying a chair or other artificial concepts, must have a large learning component, since it is implausible that humans have these capabilities entirely preprogrammed at birth. Hence our problem is not merely to explain how such recognition tasks can be implemented on a fixed number of slow, sparsely connected neurons by clever programmers. The question is even more daunting. How can slow, sparsely connected neurons program themselves to do these tasks using knowledge derived from interacting with the world?

This stark tension between the brain's substantial functionality and its severely restricted computational resources is sometimes interpreted as evidence that the problem is too difficult in the current state of knowledge and cannot be pursued fruitfully in the foreseeable future. We choose to interpret it in the opposite direction. When a problem is severely constrained it may be easier to investigate merely because the number of plausible avenues to search are few, and even the first one found may yield valuable insights.

Since the severity of the constraints is the leverage that we hope to exploit, we shall maximize it by focusing on a class of tasks to which sparsely connected, slow neurons seem the least suited. We call these *random access* tasks because the attribute that they share is that the execution of any one of them has the potential to involve any part of memory. For example, the aforementioned task of memorizing a new book title that consists of the juxtaposition of an almost arbitrary pair of words potentially requires access to any of the words the reader already knows. Associating the book with a subject, an event, or an owner requires further access to information in memory that may be almost arbitrarily unrelated to the words in the title. All the problems we shall consider have this flavor. For each one we shall give the simplest formulation that, we believe, captures at least some fundamental computational hurdle that must be overcome.

As we are focusing on tasks of a specific nature, it will be convenient for the sake of conceptual simplicity to separate the device that performs them from the devices that do not. We shall therefore hypothesize a device which we call a *neuroidal tabula rasa* (NTR) which is capable of computing the relevant random access tasks. We delegate all other tasks to various *peripheral* devices. Since the main challenge for the NTR is to overcome some fundamental quantitative barriers associated with random access tasks, our strategy will be to make the NTR as simple as possible in other respects. We shall, therefore, consider the NTR to be essentially free of preprogrammed knowledge, other than some generic algorithms needed to realize the basic random access tasks. The many cognitive processes that appear to require a large amount of preprogrammed knowledge are therefore the domain of the peripherals. These include low level sensory processing such as that of early vision, as well as, for example, the interpretation of the three dimensional world, which appears to be essential for some animals from the moment of birth. We shall also hypothesize the existence of some peripherals that are needed to mediate between the NTR and the senses. One example is an attentional mechanism that is able to focus on one part of the input scene at a time, and to present the attributes of each part to the NTR at distinct times. We do not detail how the peripherals perform their tasks. This is not inconsistent with our methodology as long as we do not require the peripherals themselves to perform random access tasks, which we do not. Also, it is not inconsistent to allow the peripherals to request the NTR to perform random access tasks for them as needed.

We note that the clear separation we impose between the NTR and the peripherals is a conceptual device adopted for the sake of ease of analysis. The question of whether the two kinds of functions are spatially integrated, which they may well be in biological systems, or separate, is not of primary relevance to our analysis here. More detailed models will need to take more detailed positions on this question in the future. The hippocampus, for example, is known to be closely involved in laying down long term memory in cortex. Although its exact role is not currently understood, it appears to interact with the cortex over a long period, perhaps several weeks, in order to perform its function. It may be thought of as a peripheral that is involved in facilitating the functions of the NTR itself, rather than in mediating or processing sensory data directly.

We shall start by giving a little relevant background in neurobiology, computer science, and cognition in each of the next three chapters. Our aim is to explain our computational methodology and to suggest that ultimately computational models should be able to bridge the gap between the neurobiological level and the level of cognitive behavior.

In Chapters 5 and 6 we shall give a definition of the *neuroidal model* of the NTR. It specifies the computational power of each component and how the components are connected in a sparse network. In subsequent chapters we specify algorithms for a variety of simple tasks that formalize memorization and inductive learning, among others. In Chapter 14 we show that the network assumptions made in Chapter 6 can be relaxed substantially, to better approximate biological reality, without the model losing the computational capabilities already demonstrated for it.

Chapters 11 to 13 deal with more complex functions, such as the representation of relations among several objects, and simple reasoning. Here computational explanations are given in terms of more complex interactions between the NTR and the peripherals. Since we do not define the workings of the peripherals to any degree of precision, the relevant descriptions in these later chapters are less complete than in the earlier ones.

Taken together the various computational mechanisms described provide a unified view of cognition. Basically there is a *cognitive substrate* of a few simple functionalities, such as memorization. When an instance of one of these functionalities is executed as a result of some interaction with the environment or of some internal process, some neural circuit may be modified. This modification can be seen as representing the acquisition of a piece of knowledge or skill. These individual acquisitions we view as giving rise or contributing to what we shall call *reflexes*. Each such reflex is simple in itself and may be viewed as a tiny amount of incremental information in the whole system. Nevertheless, when operating together these reflexes provide an effective mechanism for coping in a complex world. Intelligent behavior is then a phenomenon that arises out of the interactions of the myriad of such reflexes with each other, and with the external world.

If this view is essentially correct then it should be possible to identify the functionalities that constitute the substrate, namely those required to acquire, invoke, and maintain circuits for these reflexes. Whether or not the candidates considered in this book turn out to be correct, we believe that analysis to the level of detail we consider here, and beyond, will be required to identify the constituents of this substrate.

In conclusion, we emphasize that many of the differences between our theory and that of others derive from differences in approach or emphasis. Subtle differences in starting philosophy can lead to widely different outcomes. At some risk of oversimplification, one can attempt to characterize the various approaches according to how they deal with the three levels of function, algorithm and computational model, that we mentioned earlier.[1] For example, in the last century Boole aimed to "investigate the fundamental laws of those operations of the mind by which reasoning is performed." In other words, he believed that one can make progress by analyzing the functional level in isolation. Marr, working more recently, also believed that analysis at the functional level is crucial, but emphasized, in addition, that any mechanisms that are suggested have to be computationally feasible in the nervous system. Newell based what he called his unified theory of cognition on the mechanism of production systems at the algorithmic level, with less emphasis on specifications at the functional level or on models of computation at the neural level. The connectionist or neural network approach can be viewed as being similar except that the algorithmic level chosen is closer to the neural level.

The approach we are advocating here can be characterized as one based on "resource constrained tension between the functionality and the computational model." We believe that this constitutes the most basic computational approach. It can be summarized as consisting of three components: (a) formalizations of a range of cognitive functions, (b) a model of neural computation, and (c) the requirement that (a) be implementable on (b) within plausible resource bounds. If the range of functions is too restricted, or if the model of computation is too powerful to be realistic, then a plethora of ways of satisfying (c) exist, too many to resolve among. Experimentation designed to falsify them one by one will result in imperceptibly slow progress. On the other hand, as the range of functions is widened and the model restricted to match reality, the number of satisfying solutions diminishes toward zero only too rapidly. We believe that it is exactly within such a constrained problem formulation that it is most worthwhile to persevere in seeking a solution. As long as the functions and model are chosen with reasonable taste some such solution must exist, and if the problem is constrained sufficiently, the solution found will be close to the right one.

Suppose that on a distant planet we one day find some robots that have some very interesting and complex behavior that we would like to understand better. We could study their components and architecture much as a neurobiologist studies the brain, and construct theories of their general computational capabilities. Alternatively we could perform experiments on their behavior like a psychologist might, and construct theories of that behavior. But suppose that when these approaches are pursued separately they are not sufficient to yield the sought after insights. A third approach would be to try to imagine how one might build systems that resemble the robots in behavior, using components that resemble those that they are built from. This is basically the computational approach, in which one constructs a theory that accommodates both of the above viewpoints simultaneously and that also accounts for the computational resources such as time and hardware.

As a concrete illustration of how we pursue this approach, consider again the problem of memorizing a new book title, which we raised at the beginning of this chapter. In later chapters we shall describe a computational mechanism for realizing this task that is not inconsistent with the gross neurobiological constraints as we interpret them. At the time of writing much uncertainty still surrounds the nature and parameters of these constraints and it is therefore not clear what claims we can make regarding that particular mechanism. Nevertheless, in the broader endeavor one needs to choose somewhere to begin.

Chapter 2
Biological Constraints

2.1 Introduction

The idea that understanding the biological brain may lead to a better understanding of ourselves is a tantalizing one. It is no doubt responsible for the substantial research efforts that have been devoted to the brain over the last century, and that continue with increasing momentum. Clearly much has been achieved and substantial volumes are devoted to summarizing some of the known information.[2]

The history of this science has been punctuated with a series of striking discoveries and the development of some very powerful experimental techniques. In the 1880s Ramón y Cajal was among a number of neuroanatomists who suggested the so called neuron doctrine, which asserted that the brain consisted of cells or neurons that were discrete and physically separate from each other. Sherrington subsequently suggested the word synapse for the gaps between neurons at the points at which they came close to touching. The neuron doctrine eventually won general acceptance. In the 1920s Adrian and Zotterman recorded electrical impulses from single nerve fibers. This confirmed the view that long-range communication along nerves was electrical, but the question of how neurons communicated with each other at synapses remained open until Eccles provided evidence that this was chemical. A detailed theory of how axons produced electrical impulses, or action potentials, was finally offered by Hodgkin and Huxley in 1952. Thus, by the middle of this century, substantial progress had been made toward understanding the basic nature of the constituents of the brain.

Although at the more macroscopic or systems level an equally great amount is known, relatively less is understood. Speculations attributing various mental faculties to particular areas of the brain have been made since at least the Middle Ages. Through the study of brain-damaged patients one can attempt to more systematically associate specific parts of the brain, namely those that are damaged, with the intellectual deficits that the individuals have suffered. Associations between brain areas and functionality can also be made by electrical recordings and by recently developed nonintrusive techniques for measuring the distribution of blood flow in the brain while a subject is performing various activities. Through anatomical studies one can investigate which parts of the brain are connected to each other directly and which are not.

In spite of the wealth of knowledge that has now accumulated, the main questions of how the brain represents information and how it processes it are essentially unresolved. Indeed, there is little consensus as to how close we are to finding answers to them.

Fortunately, this book is not about the whole brain, or, for that matter, the whole mind. It is about specific kinds of memory and learning tasks that we call random access tasks. In this chapter we shall summarize those known facts about the brain that we consider to be most relevant to these functions. As will become clear, there are many simple questions to which answers are currently unavailable or uncertain. A conservative theoretician may choose to await the resolution of these before putting forward any theory. Our view here is, to the contrary, that theoretical models may even now have an important role to play. They may help to highlight the parameters that govern the basic characteristics of cortical computations, and hence encourage a greater experimental effort toward determining their values.

2.2 The Neocortex

The average adult human brain weighs about 1.4 kilograms and most of it is covered with a fairly uniform outer layer called the *cortex*. All the evidence points to the cortex as being the main seat of memory and higher brain functions, and hence the correct focus of our study. With the exception of a small part of it that is older in terms of evolutionary history, the majority of the cortex is believed to have evolved at the time of the appearance of mammals. For this reason this larger part is called the *neocortex*. For the sake of brevity, we shall often refer to it simply as the cortex.

In addition to its size and relative youth the neocortex has a further feature in terms of evolutionary history. It is a part of the brain that has grown explosively in relation to most other parts since humans evolved from early primates. We interpret this fact as an encouraging indicator for our study. It suggests that the neocortex is organized along principles that scale well with size. This scalability may have some principle underlying it that is simple enough that we may have a chance of discovering it.

Random access tasks are those that may need information from any part of memory. Hence communication over large physical distances within the brain is of central concern. Fortunately, the cortex is based on one very simple unifying principle in this regard. Long distance connections are realized in the main by one class of cells, called *pyramidal* cells. Furthermore, they form the majority of the neurons in the cortex. Therefore, as a first approximation, we can think of the cortex as being a network of these cells.

The human cortex, sometimes called the *gray matter*, is a layer of tissue having many convoluted folds, typically a little more than 2000 square centimeters in total area and a little more than 2 millimeters in average thickness. The cell bodies of the 10^{10} or so pyramidal cells reside in this thin layer. Each of these cells has a very long fiber called the *axon* that is about 0.0003 millimeters wide but up to several centimeters long, sometimes traversing the brain. The axon typically leaves the gray matter in the vicinity of its cell body, and travels within the so-called *white matter* for most of its length before reentering the gray matter at another location.

The white matter is best viewed as a cable box through which a vast amount of long distance communication is realized. Even in the technology of the remarkably thin axons, this cable box takes up a substantial volume. Figure 2.1 illustrates a section in the human brain. The large light area in the center is the white matter which is to be contrasted with the thin outside layer that is the gray matter of the neocortex. It is striking how sharply the former, which implements communication, is distinguished from the latter, which is believed to be responsible for computation and information storage. Sections across the brain at lower levels would cut parts, such as the brain stem, that are more ancient in evolutionary terms. They would show much more intricate structure and suggest special purpose functionality.

Much effort has been put into identifying the particular functions performed by each part of the cortex. This is a difficult task since many functions appear to be distributed over several areas, and most areas perform apparently very complex, possibly overlapping, functions. The areas that have proved the easiest to investigate are the motor areas and the primary sensory areas. The latter can be subdivided into primary visual, auditory, somatic sensory, and olfactory areas, which are those parts of cortex that are connected most directly with the organs of vision, hearing, touch, and smell, respectively. These sensory areas are conventionally termed *unimodal* because they appear to be influenced primarily by inputs coming from one of these senses. They are connected in turn, via pyramidal cell axons, to higher and higher areas which are ultimately *multimodal*. Cells in these latter areas respond to combinations of stimuli from two or more of the senses.

There is no consensus on the number of cortical areas that it is useful to distinguish by function. Some have put this number in the hundreds. It is perhaps in the visual area where most detailed maps of the various areas have been made. Thus the primary visual area, which contains essentially a two dimensional projection of the image on the retina, is at the back of the head and is connected by axon bundles to a series of higher and higher level regions forward of it.

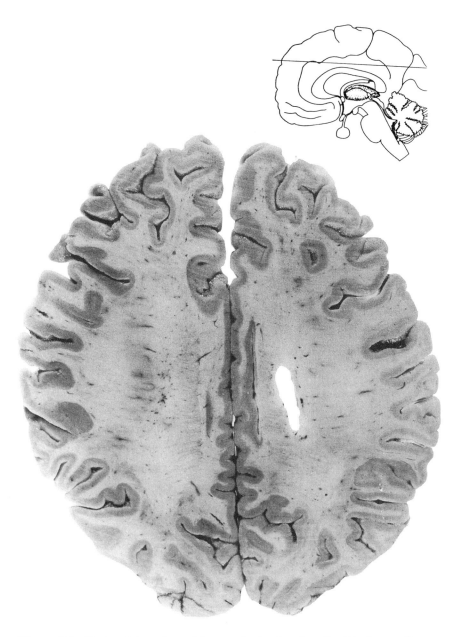

Figure 2.1. Horizontal section of human brain showing the thin layer of gray matter of the neocortex surrounding the large volume of white matter. The line in the inset shows the location in the brain of the section. From S.J. DeArmond, M.M. Fusco, and M.M. Dewey, Structure of the Human Brain, Oxford University Press 1976.

Electrical recordings show that, in the primary visual area, individual cells respond to characteristics of the scene that are local to a small region of the scene (e.g. color in a small region). At higher levels individual cells respond to what are characteristics of much larger parts of the scene. For example, in the inferotemporal cortex (IT), which is generally regarded as a unimodal visual area, there are large populations of cells that respond strongly to the sight of faces, but not to a wide variety of other stimuli. Furthermore, various subpopulations of these are selective to different features of faces. They respond in different ways depending on whether the face is frontal or in profile, or whether the eyes are visible (R. Desimone, T.D. Albright, C.G. Gross and C. Bruce 1984, C.G. Gross 1992).

While much emphasis has been placed on finding differences in function among the various parts, it is generally accepted that the physiology of the cortex is highly uniform. Experiments have shown that if the connections from the sensory organs to the visual and auditory cortices of certain young mammals are interchanged, significant functionality is retained nevertheless (M. Sur *et al.* 1988). It may be that the differences in function among the various areas of the cortex are due more to where they are connected to, rather than their intrinsic characteristics.

One pervasive phenomenon of the connections between areas of the cortex is reciprocity. Whenever there is an axon bundle going from one to the other, there is usually another going in the reverse direction. Such feedback is clearly helpful to computation of almost any nature. Whether the reciprocity is point to point (i.e. whether a connection between two cells in one direction is more likely than not to have a precise reverse connection) is currently unknown.

Another question concerns the pattern of connections between two regions. In low level visual areas, where the regions process two-dimensional projections of the visual scene, it is known that many of the connections are topographic or parallel (i.e. they connect corresponding points of the projections of the scene being viewed.) The nature of the connections in higher areas is much less well understood and may be more random.

A further issue is whether cortex should be viewed as consisting of modules at a larger scale than single neurons. Candidates for such modules have been called columns, microcolumns or assemblies. The candidate units that have been suggested vary in size from tens to tens of thousands of neurons. There is ample evidence that in the primary visual area functional units exist that are perhaps one millimeter across. The extent to which this characteristic is shared by cortical areas in general is currently unresolved.

The organization of the brain has been examined from many viewpoints. One finding is that the brains of humans are remarkably similar to those of other mammals. It has been argued that, with respect to basic components and overall organization, it is difficult to distinguish the brain of a human from that of a cow. The only striking difference is in the relative sizes of the various components. Thus the evolution of the human brain may have been primarily an enlargement of various parts in various proportions, along the lines of a uniform scalable master plan.

Within the human brain there is no question that the neocortex has enlarged enormously, as compared with the older parts of the cortex and the rest of the brain. The sizes of the various parts of the human brain have been compared with those of the most primitive mammals. One such tabulation that makes an appropriate compensation for body weight shows that the neocortex has grown most, by a factor of 180 relative to these primitive mammals, while the cerebellum associated with motion coordination has a corresponding factor of 21, and the olfactory cortex responsible for smell comes a poor last at 0.3, reflecting the relative weakness of our species in that area.[3]

The relative changes in size of the various constituent parts within the neocortex have also been significant. In mammals like rats a large fraction of the neocortex appears to be committed to the motor and primary sensory areas. In apes, and even more so for humans, this fraction is small, and the larger part of it performs activities that have proved more difficult to characterize but are conventionally associated with higher level multimodal functions.

It is possible that in addition to the massive quantitative growth, there have also been significant qualitative changes in the course of the evolution of the human brain. Identifying any specific changes has proved elusive. The most obvious functional novelty is per-

activity among several by suppressing the others.

In the dendritic tree of the postsynaptic cell the PSPs of the synapses are somehow integrated. If the overall effect is large enough then the postsynaptic cell initiates an action potential in its cell body and axon, or, in other words, *fires*. The exact mechanism that integrates the PSPs is a complex one, and the overall result may depend on the relative positions on the dendritic tree of the synapses that participate, as well as the exact times at which the various presynaptic neurons fire. If they fire simultaneously, for example, they have a larger effect than otherwise. A rough first approximation to what happens is that the various PSPs are added up and if their sum exceeds a certain threshold, say 20 millivolts, then the cell fires. We shall use this linear additive mechanism as the basis for our model. We do not have any evidence at the moment suggesting that the more complex nonlinear phenomena that have been observed within dendritic trees could be useful for implementing random access tasks, although they may have other important roles.

Action potentials appear to be stereotyped. The time profile of the impulse as it travels along any one axon appears to be essentially the same every time. Hence the information carried by a spike must be captured entirely by the instant in time at which it arrives. Beyond that, it is believed, little information is carried. This still leaves a broad range of possibilities. Information may be encoded in the average rate at which the spikes are produced, in the temporal gaps between successive spikes, or in the relative simultaneity of firings of a collection of several neurons. All three of these mechanisms may have a role, and they may be used in different ways in different parts of the brain.

The actual values of the various numerical parameters that characterize the cortex are clearly crucial to any theory of its functioning. Since our current knowledge of these parameters is incomplete, we need to use variables for these parameters to express theories and to have theories that are robust within reasonable ranges of their values. The values we give below refer to human cortex and should be regarded as very rough estimates. For some of these parameters more reliable values are known for the mouse (V. Braitenberg and A. Schüz 1991), and, perhaps fortunately for our theories, these appear less well suited to supporting generic random access functions. We also have to emphasize that in defining these parameters one runs the risk of oversimplifying the actual properties of neurons, and that methodological problems are involved in the definitions themselves (M. Abeles 1991). In general, we cannot overemphasize that all the numerical values that we quote should be regarded as provisional. In each case one can argue whether an alternative value would be more consistent with the current experimental evidence. These numerical estimates will be needed only in Chapter 14, where we shall discuss their role in neuroidal modeling. In that chapter we hope to leave no doubt that determining these parameters more precisely is of great importance.

We shall use the estimate that a typical pyramidal cell in human cortex has about 40,000 synapses on its dendrites and the same number on the axonal branchings (M. Abeles 1991). The latter number is split about evenly with about 20,000 in the local branchings, near the cell body, and 20,000 at the distant end of the axon. The dendritic tree and the local branchings are typically within perhaps 1 millimeter of the cell body. The figures we gave earlier for the area, thickness, and total pyramidal cell count suggest that there are about 20,000 such cells in each cubic millimeter of cortex. Hence two cells within 1 millimeter of each other have a significant chance of synapsing with each other if the dendritic trees and axonal branchings of each of them synapse randomly in each others' vicinity. This is simply because the 20,000 synapses on the local axonal branchings of one neuron have nowhere else to go than to spread themselves roughly evenly among the 20,000 or so neurons in their vicinity, assuming for now that they spread throughout a volume of one cubic millimeter.

The value of a PSP is modifiable and depends on the history of the firings of the two cells involved. One process of modification is called *long-term potentiation* or LTP, and has been widely investigated. It is believed that the dominant vehicle for storing learned information in the cortex is in the values of the PSPs, which correspond to *weights* in the neuroid model. It is also thought that in humans no new neurons are formed after birth, or even after the first four months of gestation. The growth of new connections is less well understood, but it is not currently believed to provide the primary mechanism of learning.

It appears that the PSPs do not constitute the only mechanism available to neurons for storing information that is dependent on the previous history of activity. A wide variety of others have also been investigated (C.D. Woody *et al.* 1988). Perhaps the simplest such example of history dependent behavior is that of a *refractory* period following an action potential. During such a period the cell will not fire even if the stimulation at the synapses would be sufficient to make it do so normally. In the neuroid model memory mechanisms that are associated with the whole neuron, rather than with individual synapses, will be modeled by *states*.

A fundamental open question, which relates to the style of computation and information representation used in the cortex, is the number of presynaptic neurons that need to fire in the course of a typical natural computation in order to make a postsynaptic neuron fire. It is believed that a total contribution of about 20mV is needed to make a neuron fire. Also, it seems reasonable to assume that the *average* contribution of any one synapse should be a very small fraction of this, less than 0.1mV. (This is because cortical neurons are known to fire apparently randomly in the background at rates in the range of perhaps 0.5 to 10 times a second. Assuming a rate of 5 here, it follows that in any millisecond about 200 of the 40,000 dendritic synapses will contribute a PSP. The sum of these 200 PSPs should not be enough to make the postsynaptic neuron fire.) Such a low average PSP value is well confirmed by the available experimental evidence. The significant open question is whether among its many dendritic synapses, a neuron has at least a few with significantly higher than average PSPs. Some evidence has been found that this is indeed the case. In rat cortex PSPs above 2mV have been observed (A.M. Thomson *et al.* 1988, A. Mason *et al.* 1991, A.M. Thomson *et al.* 1993) which is much higher than the corresponding average values. If higher PSPs exist but only on a few synapses for each neuron, (e.g. less than 0.1% of synapses), and only in brain tissue that performs random access tasks, then it may yet prove difficult to confirm their existence. Further complications arise from the difficulty of distinguishing between single versus multiple synapses. Also the values of the

PSP contributed by a synapse seem to vary randomly on different occasions. The idealized weight is at best some average value.

The strongest evidence that a single neuron may have a large influence on its neighbors comes from correlation studies (D.H. Perkel *et al.* 1967, M. Abeles 1991). Electrical recordings are made simultaneously in or near two neurons. These signals are analyzed statistically. If activity at the first correlates highly with activity at a second, say one millisecond later, then it is a plausible interpretation that these two neurons are directly connected. A different pattern of correlation is interpreted as signifying that the two neurons are not directly connected, but are connected to a common presynaptic neuron. The majority of experiments confirm that the typical influence of a neuron on another is very small. In a few cases stronger influences have been found. For example, one experiment on the inferotemporal cortex (P.M. Gochin *et al.* 1991) suggests according to this interpretation that there exist some pairs of cells in which one controls the behavior of the other quite strongly. The spiking of one is highly correlated with the spiking of the other at a time immediately before.

The quantitative issues raised in the two paragraphs above are crucial, we believe, in any analysis of the capabilities of cortex for performing random access tasks. In Chapter 14 we introduce a parameter α, which is defined there as the minimum number of synaptic weights of a model neuron that can sum up to the threshold needed to cause it to fire. Relating this parameter to real biological neurons raises several issues. For example, it is conceivable that in natural computations a significant fraction of the 20mV threshold that needs to be overcome by a neuron is contributed by the random spontaneous firing of its neighbors, in which case the purposeful activity would not need to contribute as much. An effective value of $\alpha = 1$ is the one that would support our algorithms most easily. However, at present there is no direct evidence to support values of α less than about 5. For this reason we shall expend some effort in Chapter 14 to show that our algorithms can be supported by larger values of α, such as 5. There is a fundamental computational hurdle that needs to be overcome if we are to work with such higher values of α. The basic problem is that in order to make a neuron fire, either several presynaptic neurons now have to be coordinated or, alternatively, there has to be one presynaptic neuron that makes several synapses with the same neuron. Perhaps the only previous work that addresses this issue in detail is that of Abeles. He gives a method of performing a certain kind of communication within local cortical circuits (i.e. neurons densely connected with each other) using what he calls *synfire chains* (M. Abeles 1991). The problems we need to solve in our formulation are more constrained, not only because our random access tasks impose a more onerous burden on the network, but also because our network models the sparser connectivity realized by the long range axons.

There are three further important aspects of the cortex that models may need to take into account. First, while pyramidal cells are the majority of neurons in cortex, they are not the only ones. In particular, there are cells that are inhibitory. Their firing has the effect that postsynaptic neurons synapsing with them are restrained from firing. Fortunately, for the modeler, these neurons do not have long distance connections. Thus if one cell A is to inhibit the firing of a distant cell B, it would appear that A would need to excite an inhibitory neuron C in the vicinity of neuron B.

A second aspect is that the cortex is conventionally divided into six parallel *layers*, layer I being at the surface and layer VI closest to the white matter. The layers can be distinguished according to the types of cells found in them, the destinations of any axons that emanate from them, as well as the distal axonal branchings that terminate in them.

Lastly, and as mentioned previously, numerous authors have emphasized that the cortex is not homogeneous laterally (V. Mountcastle 1979, J. Szentágothai 1978). It is believed that the cortex can be viewed as consisting of units, variously called *columns*, or *assemblies*, that have the shape of cylinders cutting through perpendicular to the six layers. They are believed to be interconnected more richly internally than they are to neurons in neighboring units. There is some disagreement, however, about the details of the size and nature of these units.

In addition to α, the models introduced in Chapter 14 will also assume particular values for some other parameters, namely ρ, the expected number of synapses between two neurons in a certain proximity, and χ, the average number of synapses on the local and distal axonal branches of a neuron, for which we have already quoted estimates. We note that determining these parameters accurately presents numerous difficulties. An obvious one is that experiments that count neurons and synapses will yield some average value for χ, for example, over all neurons, while what we seek is its value for pyramidal cells. Furthermore not only are each of these parameters difficult to determine experimentally, but also, any direct measurement may be difficult to interpret, since the *effective* values of these parameters in the actual neural computation may be influenced by additional factors. For example, the effective values of all three of these parameters may relate to synapses formed by pairs of neurons in a specific pair of layers. If a parameter such as ρ is measured over all pairs of neurons then the value obtained may be rather lower than the effective value it takes in the actual layered arrangement. If one accepts the six layer classification, one can ask thirty distinct questions about how the local axonal branching of the neurons in each layer synapse with neurons in each of the others. The answers to each of these questions may provide the values of ρ that are the effective ones in the various computational behaviors. Some of these values may be rather higher than the overall average value.

Chapter 3
Computational Laws

3.1 Introduction

Although computation has had a central role in mathematics from the beginning it was not studied systematically until the recent advent of digital computers, which made large scale computations possible. A new discipline has arisen which has come to be called computer science. From this a broad perspective has emerged that invites comparisons with the physical sciences. In particular it now appears that the world of computation can be compared with the physical world not only in the richness and variety of phenomena that can be observed, but also in the existence of underlying laws that govern what is and what is not feasible to do.

It is possible to view the laws of physics as negative statements of what cannot be done. The law of energy conservation, for example, states that energy cannot be created or destroyed. The equation $E = mc^2$ does not say how matter can be converted into energy but does limit the quantitative nature of any such conversion that does take place. As we know, an understanding of such negative constraints can and does lead to positive consequences. By ruling out a myriad of fruitless paths it helps channel the search for how to achieve some desired result in the physical world along constructive lines.

Computational laws may be interpreted as having an essentially similar negative nature. They capture the limits beyond which computations are impossible. As Turing showed in an epoch-making paper, some well defined problems do not have computational solutions at all. Others cannot be solved with less than a certain amount of resources, such as of computational steps or storage space.[4] A better understanding of these limitations can be expected to and does yield positive benefits, just as it does in the physical world. In the current context we are concerned with understanding the limits that brain-like models impose on the computation of random access tasks, so that we may also better understand the possibilities.

Central to our notion of a computational law is the distinction between a computational problem and an algorithm. The former is a statement of *what* is to be accomplished while the latter describes *how* it can be done. For example, consider the familiar problem of multiplying two large numbers. Suppose that we agree to represent numbers in conventional decimal notation. Here the task that has to be performed is unambiguously clear. There are, nevertheless, any number of algorithms or methods for performing it, even if we fix the ground rules. Suppose, for example, that we agree that any algorithm has to consist of atomic steps that operate on a pair of single digits. Long multiplication as universally taught in elementary schools is just one option. It is now known, but only since surprisingly recently, that long multiplication is by no means the best method for large numbers. If the numbers to be multiplied are each n digits long, then this conventional method would multiply each digit of the second number by the whole of the first number taking about $2n$ steps for each such digit. The total cost of doing this for all n digits is therefore $2n^2$, and a similar cost is incurred in finally adding up the n results obtained. There appears to be no record of anyone knowing of a better algorithm prior to 1962. In that year Karatsuba and Ofman published an algorithm in which the number of steps grows as $n^{1.6}$ rather than n^2. It improves on the conventional method even for moderate values of n, and does so more and more dramatically as n grows.[5] Since then even faster methods have been discovered that reduce the growth of the runtime as n grows, to almost linear in n.

A more elementary example is that of division. Consider the problem of dividing a number consisting of $2n$ decimal digits by another of n decimal digits. One method of doing this is to repeatedly subtract the second number from the first, and count how many times this can be done before zero is reached. This algorithm is a correct one, but takes exponential time, about 10^n steps. Compared to this the familiar long division algorithm looks good, requiring only about n^2 steps, and we should be thankful both that it exists and that someone discovered it.

An algorithm cannot be described totally in the abstract. One has to assume some specific model of computation. Such a model specifies the basic individual steps that can be performed that are to be regarded as atomic and not broken down into smaller parts. In the previous paragraphs the bounds given on the number of steps taken apply for several models. One such model is that of a human doing the calculation with pencil and paper and counting each basic arithmetic operation on a pair of decimal digits as one operation, with all other actions, such as writing digits, counted as having no additional cost.

The usefulness of computational laws rests on the observation that models of computation can be very *robust*. Once a model is powerful enough to capture certain computational phenomena, often many changes in the details of its specification can be made without changing its computational power. For the multiplication problem the n^2 and $n^{1.6}$ step algorithms mentioned keep their respective behaviors for a very broad range of models that include not only pencil and paper calculations, but also digital computers programmed in any of the standard high level languages. Small differences that do arise are often in the constants that multiply the growth function. For example, long multiplication can be charged as $2n^2$ or $4n^2$ steps, or whatever, depending on the details of the model. For this reason these constants are often suppressed by means of the "big O" notation. Thus $O(n^{1.6})$ means "at most $kn^{1.6}$ for some constant k independent of n." Although this notation can be abused by using it to hide enormous constants k, its value is in highlighting the order of growth, which is usually the most significant component in practice.

The discovery of robustness phenomena for computational models was among the first significant achievements in computer science. In the 1930s the work of Turing, Church, and others established that several superficially dissimilar models of computation had identical computational power. The functions that could be computed on any one were the same as those that could be computed on any other. This early study led to the development in the 1960s of the theory of computational complexity, in which emphasis shifted from the issue of what can be computed at all to that of what can be computed efficiently. Here again strong robustness phenomena have been established in many directions.

By analogy with these previous success stories we shall hypothesize that sufficient robustness exists around the neural model that we shall develop here, to give the study validity. Certainly several simple mutations to the model can be shown to leave its power invariant. Any study based on simplified models requires some such robustness to justify it. We do not want the results to be artifices of the arbitrary choices made in the definitions.

In the light of this introduction we can formulate the nature of a typical *computational law* as follows. First there have to be defined both a model of computation, and also computational problem or task. The law then states the ultimate limitations that any algorithm on that model for that task has to confront. The limitations may be in terms of any of several criteria, such as the number of steps required, the amount of storage spaced used, or the accuracy or reliability of the solution.

Once a model of computation is accepted as useful, the limitations to computing a well-defined task on it can be formulated as mathematical questions. Hence, in principle, these limitations are resolvable with mathematical rigor. In this sense a computational law can be placed on firmer foundations than a physical law. Of course, there always remains the informal aspect of what constitutes a useful model of computation. This aspect is also present, however, in physical laws, in which the computational models correspond to irreducible concepts such as, for example, force or energy. Physical laws about force or energy are of interest only because they turn out to relate to some aspect of reality, a fact which is not a law itself.

While in principle computational laws can be given more solid foundation than physical ones, this extra rigor has not been achieved to date except in very restricted domains. For significant areas of computation our current knowledge of what can be computed is expressible only in the form of hypotheses which, much like physical laws, are subject to future falsification. A prime example is the hypothesis, first enunciated by Cook, that none of the so-called "NP-complete" problems have efficient algorithms.[6] The weight of evidence for or against such conjectures is impossible to evaluate except in terms of the human effort that has been expended in attempts at resolving them. If many people have tried and failed to find an algorithm with a specified efficiency, then in human terms there is more evidence that no such algorithm exists than would be the case if no-one had sought to find one. Also, as long as we do not mind being wrong, there is no reason for not hypothesizing that the best algorithm we know at any time is the best possible. Indeed, this methodology seems the most promising approach at the moment for pinning down the ultimate limitations in resources needed for performing significant computations. It is in this spirit that the algorithms for the neuroidal model, described in later sections of this book, are intended. We do hope, of course, that eventually mathematical techniques will become available for establishing these laws once and for all, but for the moment we apparently must follow this less rigorous strategy, as is currently done in most other areas of computing.

3.2   Three Sources of Complexity

It is useful to distinguish three distinct sources of computational difficulty that the mechanisms of the brain have to overcome. These are computational complexity, descriptional complexity, and learning. Each of these is studied as a subfield of computer science.[7] We shall endeavor here to avoid using technical results from any of these areas. It is important, however, to call attention to the phenomena at their respective centers, since there lie the impediments which our algorithms, or, we believe, any other computational theory of the brain, will need to confront. It is the reality of these impediments that underpins our general methodology, since it guarantees that there is some virtue in exhibiting mechanisms that overcome them, whether or not these mechanisms are actually the same as those used in the brain.

The first, *computational complexity*, is the one from which we drew the examples in the previous section. The phenomenon associated with it is that a computational problem may have substantial intrinsic computational difficulty. A certain amount of resources, such as number of steps or storage space, at the minimum, will be required for the computation. The resource bounds in the brain, as currently understood, are seriously limited. A fixed number of about 10^{10} neurons need to suffice for a lifetime. Significant recognition tasks can be performed in about 100 milliseconds, which allows for relatively few successive steps of cortical firings. Clearly, any functionalities attributed to the brain have to be such that these computational resources are demonstrably sufficient.

The second, *descriptional complexity*, is concerned with the fact that for any one function, while there may be several different programs for computing it, these programs will require some minimal length of description. To make this concrete one needs first to fix an appropriate language for specifying a program, perhaps a standard programming language, a theoretical model such as a Turing machine, or our neuroidal model, and then agree on how to measure program length in terms of it. Now, all the empirical evidence with computer programs suggests that to achieve any significant functionality appropriate for complex situations, one needs long programs. This is consistent, for example, with the experience of most computer users. The moral here is that, when we seek to find out exactly what the brain does, the target of the search should be of the order of the complexity of a set of programs rather than a single equation as found in physics. Even if the programs have some unifying underlying principles, which we certainly hope that they do, in their totality they will be long. The interactions among them during execution may be even more difficult to describe since, in general, timing issues in distributed and parallel systems may be very complex. Furthermore, this descriptional complexity does not arise only because much new information is acquired by the brain through learning. We are suggesting that the computational mechanisms that underlie the basic functions of memory, learning, and recall already are of substantial complexity. The evidence from biology is certainly consistent with this view. It has been estimated that fully a third of the mammalian genome is dedicated exclusively to the functions of the brain (J.G. Sutcliffe 1988). Reconciling ourselves to this source of complexity may be, in itself, a significant step forward.

Finally, there is a third source of difficulty, that of the inherent complexity of *learning*. The various programs in our brains are either present at birth, a result of evolution, or have been learned during life, a result of interactions with the world. Some may be the result of some combination of the two. Their totality is, as we explained above, most probably of substantial descriptive complexity. In order to have any chance of uncovering their nature it seems essential that we take some intellectual view of how these programs got into the brain in the first place. How do these programs relate to the interactions with the world that produced them during evolution or learning? How do they relate to the world in which they are to perform effectively? In the case of learning, these issues are addressed in the field of computational learning theory. The phenomenon that needs a quantitative explanation is, essentially, the following. How can a system, with limited computational resources and exposure only to a moderate number of situations, acquire programs that are going to be effective and robust in dealing with new situations not previously seen? Since, as we believe, there is a very large amount of information in the brain that is acquired by learning, some position on the nature of this process has to be taken. We shall return to this issue in later chapters.

Chapter 4
Cognitive Functions

4.1 Introduction

In the previous two chapters we considered the possible contributions that the fields of neuroscience and computer science could make to an understanding of cortical computations. In this chapter we consider the third major viewpoint, that of cognitive psychology. The potential contribution of this field is to provide definitions or specifications of the behavior exhibited by the brain. Once accurate specifications are available, and provided suitable neural models have been abstracted from neuroscience, there remains only the problem of discovering the algorithms.

It appears that among these three aspects the problem of specifying the cognitive functions presents the most formidable difficulties. For a complicated device such as the brain, it is difficult both to describe the total behavior in its full complexity, as well as to decompose it into simpler constituents. There exist theoretical results that show that even relatively simple computational mechanisms can result in behaviors that are so complex that a description of the mechanism or of the constituent parts of the behavior cannot be recovered feasibly from observations of the behavior itself.[8] Hence we have to hope that there is some *substrate* of simple functions on which human cognition is built, and which we can discover by, dare we say, inspired guesswork.

Where should we look to find candidates for the functions of this cognitive substrate? One approach is to study the results of experiments in cognitive psychology and to use them to identify the basic functionalities that underly behavior. In practice this appears to be difficult to do. It may be that the difficulties mentioned in the previous paragraph account for this impediment. In this volume we follow what is essentially a reverse approach that starts with some simple functionalities and then traces the implications that follow from them.

Attempts to isolate mental faculties and to describe them in explicit mathematical terms have often met with serious difficulties. More than three centuries ago Leibniz made exactly such an attempt to capture the nature of thought processes. At the age of nineteen he wrote a dissertation *De Arte Combinatoria* on this topic, and he returned to it several times later in his life with less success than he would have wished. Little further progress was made until the 1850s when Boole published his *Laws of Thought*, which fulfilled some of Leibniz's original aims. Boole constructed a mathematical system, now called Boolean algebra, that was explicitly motivated by questions of cognition. In the next section we shall discuss his system and explain why it is a useful starting point for us. In subsequent sections we shall explain and justify how we selected the actual functions that we chose to study for implementation on the neuroidal model. Although on the surface these functions look simplistic and even impoverished, it turns out that their implementation on neuroids suggests a rich world of computational and cognitive phenomena. As we explain in the final section §4.5, these phenomena are amongst those of greatest current concern to experimental cognitive psychologists.

4.2   Boolean Functions

Boolean algebra formally resembles traditional algebra but is intended to be applicable in a different domain. The basic difference is that a variable x in Boolean algebra can take values "true" or "false" rather than numerical values such as 2 or 3.6. As a consequence the meaning of the operations needs to be different, since the conventional arithmetic operations, such as multiplication, no longer make sense. Finally, the manipulations that it is valid to perform on equations must also be different.

The intention of a Boolean variable is to represent a proposition. Thus x can stand for "it is raining" and y for "it is cold." The intention of a Boolean operation is to create new propositions from old ones. For example, the operation "and" denoted by "$\wedge$" (or sometimes by ".") can be used to create the new proposition "$x \wedge y$" which in this instance would denote the proposition "it is raining and it is cold." In a similar way "or" is a useful operation, denoted by $\vee$, (or sometimes +) and $x \vee y$ would denote the proposition "it is raining or it is cold or possibly both." A third operation is "not" and is denoted by $\neg$. Thus $\neg y$ would denote that "it is not cold." While $\wedge$ and $\vee$ are binary operations, i.e. having two arguments, $\neg$ is unary and has just one. Boolean algebra is concerned with the laws under which expressions formed by Boolean operations can be manipulated. For example, $(\neg x) \wedge (\neg y)$ is equivalent to $\neg(x \vee y)$, as it should be under the semantics just described for these operations. By equivalent we mean that for any combination of the two possible truth values for x and the two possible truth values for y, the truth value of the two expressions will be the same.

It has been noted with surprise that Boolean algebra was not discovered earlier. For example, Leibniz discovered the differential calculus apparently with ease, while his comparatively greater efforts towards formalizing thought processes apparently failed to uncover Boolean algebra[9]. One possible explanation for this is that while the differential calculus is an outstandingly clear theory of continuous processes, Boolean algebra is so incomplete in explaining thought that it was simply rejected by those who might have considered it.

Perhaps the most important aspect of Boole's work is that it provides a model of cognition in which variables can take on only a discrete choice of values, in particular "true" or "false", rather than an unbounded choice. The merits of this central idea, that cognition should be modeled in terms of discrete mathematics, can be debated, but in may ways it has withstood the test of time remarkably well and it permeates current thinking. There is now much accumulated experience with handling human knowledge in large quantities, whether in print or in computers. The representation used is almost invariably discrete. When one opens an encyclopaedia one sees discrete words, not arrays of dials or decimal numbers.

Here we shall adopt this view that discrete representations play a fundamental part in the substrate of principles on which human cognition is built. For searching, updating, and structuring knowledge as our algorithms in later chapters attempt to do, we know of no competitive alternative approach. Clearly, by itself Boolean algebra is at best an incomplete theory of cognition when compared with successful theories in some other fields, such as physics. However, rejecting it for that reason would be a mistake. The reality may be that the true "laws of thought" are of much greater descriptional complexity than the basic laws of the physical sciences, and Boolean algebra captures only a part of it.

For these reasons we shall formulate the various cognitive tasks in a Boolean framework in the first instance. As will be clear later, the actual functions implemented by our algorithms, even the simplest ones, will have more structure, being "softer around the edges." This softening is a consequence of the style of implementation we use, and of the random properties assumed of the interconnection pattern. In addition, we will introduce further internal structure by allowing relations, for example, as constituent parts of the Boolean functions.

The most basic cognitive tasks we consider are those of *recognition*. These will be implemented by circuits that evaluate the corresponding Boolean function or *predicate*. For some set of input variables — say, x, y, z — the circuit will output "true" or "false" according to the value the function takes when supplied with truth values for $x, y,$ and z. The input variables may be themselves the outputs of Boolean functions and may express something complicated. Thus the function $w(x, y) = x \wedge y$ may express "it is a nasty day." In general we consider that some *scene* is presented to the system, and the truth of the predicates are evaluated by the system for that scene.

A central question is to determine which classes of Boolean

functions are most appropriate for modeling how knowledge is represented in the brain. In particular, we ask the incremental version of this question. If one new piece of knowledge is learned, what is the appropriate class of representations from which the act of learning selects? Classes that are too general become unrealistic if no plausible mechanism can be found for learning instances of them. Classes that are too restrictive are unrealistic if they cannot express significant fragments of real world knowledge.

A simple but unavoidable class is that of *conjunctions*. If variables, $x_1, \cdots, x_{100}$, say, are available, a conjunction is the "and" of any subset of them, such as

$$x_3 \wedge x_7 \wedge x_{19}.$$

These are sometimes called Aristotelean concepts and have been discussed at length by philosophers and psychologists. The truth of every variable that appears in the conjunction is both necessary and sufficient for the conjunction to be true.

Conjunctions seem relevant to representing single instances of objects or events. They express the conjunction of all the relevant attributes. Thus, the notion of "yesterday's lunchtime companion" contains simultaneously the attributes, yesterday, lunchtime, and companion, and it seems difficult to avoid expressing it as some kind of conjunction. For this reason we shall accommodate conjunctions centrally. On the other hand, it is clear and has been extensively argued that more general concepts require richer representation classes. Wittgenstein argued that the notion of a "game" has no attributes that are both necessary and sufficient. For example, not every game is won or lost, not every game is played by two people, etc. For these reasons we need to go to more general representations. A most attractive generalization is *disjunctive normal form*, abbreviated usually to DNF. A DNF expression is a disjunction of conjunctions, such as

$$(x_1 \wedge x_5 \wedge x_7) \quad \vee \quad (x_2 \wedge x_5 \wedge x_7) \quad \vee \quad (x_2 \wedge x_4 \wedge x_8).$$

DNF can express concepts that have several distinct varieties of typical members. If we are allowed to write one conjunction to characterize one-person games and another to cover two-person games, we would get a better characterization of games than is possible with a single conjunction.

One important point is that, in all of this discussion of conjunctions and DNF, we could have replaced any single variable, such as x_3, by its negation $\neg x_3$. Indeed these classes are defined conventionally to allow negations. In this text, however, we will avoid using negations as much as we can. The firing of a neuroid will correspond usually to a variable x being true. If we need a variable to represent the opposite, we can associate a separate neuroid with a new variable, say y, that will have positive value when the neuroid fires and is logically equivalent to $\neg x$. However, on a few occasions it will be necessary to face the issue of negation or inhibition explicitly.

4.3 Learning

Learning phenomena have been classified and categorized in numerous ways. Here we shall adopt two dichotomies that appear to be fundamental in any context in which we wish to describe explicit mechanisms of learning.

The first dichotomy is between *memorization* and *inductive learning*. The first of these is simply the storage of some information that is explicitly provided or internally deduced. The information memorized may be, for example, the spelling of a word, or it may relate to the appearance of a person, the description of an event, or the result of a logical deduction. The second notion, inductive learning, we define essentially negatively, as any kind of information gathering where the information acquired is not explicitly given or necessarily implied by that which is explicitly given. The common characteristic that the phenomena of inductive learning have is that some form of generalization is involved that is not dictated unquestionably by the evidence. When learning to recognize chairs from some examples, we acquire a capability that is somehow more general than the ability merely to recognize the particular examples of chairs that we have seen. The riddle that inductive learning presents is that, while it seems difficult to defend it intellectually, it does appear to work in the real world.

In conventional computers memorization is a trivial operation, and for this reason in the field of machine learning it is often not regarded as learning at all. In the neural context, however, it raises very challenging computational problems that are too easily underestimated. The challenge of modeling inductive learning is, of course, even greater. Learning involving generalization poses fundamental philosophical questions as to its very nature that memorization does not. Nevertheless, as mentioned in the previous chapter, we need to take a concrete view of it in any theory of cognition. The viewpoint we shall adopt is that of computational learning theory and we shall discuss it further in Chapter 9. We note that essentially all knowledge acquisition phenomena, except those modeled as memorization, appear to have some component of generalization. For this reason we shall use the word *learning* as an abbreviation for inductive learning.[10]

The second dichotomy is between *supervised* and *unsupervised* learning. Consider the process in which the learner is presented with some examples. In the case of supervised learning, the information describing each example is accompanied by information of a second kind called the *labeling*. The labeling could be provided by a teacher or deduced by some internal process by the learner. When learning the concept of a chair, we will be presented with a sequence of examples each labeled as "chair" or "not chair". In unsupervised learning, on the other hand, information describing the examples alone is presented, with no additional commentary.

These two dichotomies can be used in several complementary ways. In one case we could learn the sound of the word by unsupervised memorization, and then learn its meaning from a sequence of labeled examples in supervised inductive mode. In another case, we can first learn in unsupervised inductive mode conjunctions of attributes that often occur together in the world. In supervised inductive mode we can then learn a more complex Boolean function than we could have otherwise, by treating these conjunctions as equal citizens with previously available single attributes.

More specifically, we shall use these two dichotomies to describe four modes of learning. We can characterize these by further examples of them as follows: *unsupervised memorization* is appropriate for memorizing the sound or spelling of a word about which we have no previous knowledge; *supervised memorization* will be appropriate for associating the face of a person with a name; *supervised inductive learning* is appropriate for learning a concept or category such as "chair" or "good" from examples labeled as positive or negative instances of the given category; unsupervised inductive learning is typically to do with spotting combinations of events or attributes that occur together unexpectedly often, and we shall refer to this as *correlational learning*.

In each case the aim of the learning process will be to set up a neuroidal circuit that, when given inputs subsequent to the learning experience, will categorize the input according to the required function. Thus the aim of memorization is to set up a circuit such that repeating the same input will result in recognition. The aim of inductive learning is similar, except that recognition is expected for a broader class of inputs.

It is worthwhile to relate these dichotomies to those made in the psychology literature. The most relevant such distinction is between "declarative" and "nondeclarative" memory. The former relates particularly to facts and events, where what is remembered is accessible to one's conscious recollection. The latter relates to skills and habits, where the details of the skill (e.g. how to ride a bicycle) are not accessible to one's consciousness. This declarative/nondeclarative distinction is similar to our memorization/inductive learning distinction, provided the recognition of general concepts, such as chairs, is considered nondeclarative. Many other related distinctions have been considered also (E. Tulving 1983, L.R. Squire 1992).

4.4   The Nature of Concepts

The nature of a concept, sometimes called a category or a universal, has been a central subject of philosophical speculation. When we describe something as a "game" or a "chair" we mean that it belongs to a certain class. The basic question is to delimit more carefully the nature of the classes that humans employ as categories.[11]

Saying that concepts correspond to Boolean functions is a step toward taking a position on this issue but, as we shall argue, only a small one. At one level it is a statement that is trivially true, since if we digitize the input presented and define the Boolean function of these inputs as having value true if the human characterizes the input as "chair" and false otherwise, then we have a Boolean function at least for one person's notion of chair.

For our current purposes the better question to ask is whether there exist more detailed models of concepts that are more useful. One major problem area was discussed in previous sections: namely, which classes of computationally tractable knowledge representations are sufficient to represent human concepts? Is it Boolean conjunctions, or disjunctive normal form, or something else? Psychological experiments do yield some clues here. It has been found (S.J. Thorpe *et al.* 1989), for example, that artificial concepts made up by combining elementary concepts by "and" or "or" are much easier to learn by humans than those composed by the "exclusive-or" connective (i.e. one or other but not both.) Attempts by workers in artificial intelligence to describe natural concepts formally yield further clues. Notation for expressing relationships among constituent parts of the examined information appears to be indispensable. We shall examine this latter issue in detail in Chapter 11.

Boolean concepts may have internal structure in several additional ways. The *exemplar* theory suggests that our representation of a chair consists of descriptions of particular chairs that we have seen. On seeing a new object we compare it with these exemplars and see whether it is sufficiently similar to at least one of them. The unanswered problem here is to describe measures of similarity that work for the whole range of human concepts. Related theories suggest that concepts are *graded*. Some chairs are classified by human subjects as more typical than others. Furthermore, this subjective measure of typicality correlates with more objective ones, such as reaction times measured when subjects are asked to categorize objects. A third theory claims that there are a number of attributes that are each positive indicators of chairhood, such as having four legs and being suitable for sitting on. A chair is then anything which satisfies at least a certain number of these.

The view of concepts that emerges from the representation and algorithms described in our later chapters does have something in common with all these theories, but is more complex than any one of them. Perhaps the main difference is that a concept is no longer unitary. For example, although any individual's notion of "France" is at some level a single Boolean function, it is more useful to view it as the interaction of perhaps a large number of functions variously acquired by memorization or inductive modes. Thus we may have in our mind, as a prototype, a map from an atlas we had in childhood. In addition, we can clearly recognize outlines that are close enough. This ability we consider here to be acquired by inductive learning. In addition, there may be many items of information acquired by memorization that are associated with these inductively learned functions, in this case, perhaps, the names and locations of cities. If we see an outline map that resembles France and a dot inside labeled Rome, the immediate computational reaction is probably not usefully viewed as the evaluation of the unitary concept of France, but rather as the recognition of a number of distinct predicates. These may turn out to be inconsistent with each other, in which case we will need to resolve amongst them by other means such as, for example, the application of further functions that depend also on the outputs of the functions that were applied first.

One consequence of our choice of knowledge representation is that even though we treat the functions learned as Boolean in the first instance, they do come to have fuzzier edges than are conventionally associated with Boolean functions. The main reason for this is that whenever we require in the implementation that particular neuroids be connected, this is ensured only with high probability. For example, when we learn conjunctive concepts the representation has some of the flavor of the last psychological theory mentioned, where instead of the requirement that all of a set of necessary and sufficient attributes are present, the presence of a sufficiently high fraction of a set of confirmatory features is enough. The fact that neuroids act as threshold functions also contributes to the nature of the final representation realized.

We emphasize that the choice of knowledge representation is a most central issue in analyzing models of neural computation. Instances of the representation have to be expressible succinctly in the neural model. Also, there has to be a plausible account of how these structures can be learned. As we shall see later, the representation that emerges from our model is a complex one, influenced by our view of the discrete nature of concepts, by the consequences of the properties of random graphs, as well as by the algorithmic properties of the neuroidal model.

4.5   Experimental Psychology

Over the last century a large body of experimental data has been collected regarding the cognitive performance of humans and other species under various laboratory conditions. The phenomena studied have included memory, learning, attention, as well as numerous others.[12] Many of these results are robust and reproducible in exactly the same way as are experimental results in the physical sciences. The most glaring difference is that, as compared with, say, physics, no global theories have emerged that account for comparably broad ranges of phenomena and can make comparable predictions.

Our view here is that a successful search for such global theories, for example, of the cognitive substrate, will need to be theory driven. One needs to start with theories that have the potential to be global. Experiments can then resolve among the candidates. The neuroidal formulation described in the next several chapters is intended to facilitate a range of just such theories. The development of this formulation was influenced little, however, by current trends in experimental psychology. We made no explicit attempt to ensure that the model fit psychological facts, and for this reason we have left discussion of the latter to this last section.

In developing the model we found to our surprise that several mechanisms that we introduced to overcome computational impediments corresponded closely to notions that already had wide acceptance in cognitive psychology. These include, for example, attentional mechanisms, imagery, and serial processing of multi-object inputs. We shall devote the remainder of this section to a review of some of these existing connections, and to some brief comments about where further connections might be usefully sought.

We start with a very basic question concerning scenes that contain more than one object. The propositional predicates of Boolean algebra can be applied to a whole scene, but how are they to be applied to parts of a *multi-object scene*? For example, in the normal interpretation when a picture is being described, the predicates blue or green would describe the whole picture. How are we to treat the case where the picture contains several distinguishable parts, one of which is blue and another green? The question of how humans deal with this situation has been the subject of careful investigation (A. Treisman and G. Gelade 1980). In a typical experiment a human subject is presented with pictures that each contain a number of figures, such as green triangles, red squares, etc. The subject is asked various questions and the time required for answering them reliably is measured. A typical finding is that when asked whether the picture contains an object having a single attribute, such as being green or being a triangle, the time taken is independent of the number of objects. The interpretation of this is that processing of all the objects is carried out in parallel, and perhaps a global propositional predicate, "there is a triangular object," is evaluated for the whole scene, in time independent of the complexity of the scene. In contrast, if an object having a conjunction of two attributes is sought, such as a green triangle, then the time taken increases linearly with the number of objects, suggesting that the subject is, at some level, processing the objects in sequence. This sequential strategy is exactly the solution we adopt in the neuroidal model in §11 to deal with multi-object scenes. It is the simplest general computational mechanism we could find for this task. We did not attempt to make any fit with psychological data, which, as it happens, turn out to be rather complex. (K. Nakayama and G.H. Silverman 1986, A. Treisman 1988).

One closely related issue is that of *attention*. Conjunctions of attributes in a part of a scene are typically not noticed if they have not been attended to. In experiments of the kind just described subjects sometimes report having seen a green triangle and a red square when they have been presented with a red triangle and a green square for too short a time to attend to them separately. This phenomenon is called *illusory conjunctions.* In everyday life we fail to recall conjunctions that we have no motivation to have noticed. Few of us can recall which letters are associated with which numbers on a telephone dial, although we have been exposed to these inputs innumerable times[13]. Even single attributes need to be attended to to be remembered. People find it difficult to recall, for example, the direction in which the head faces on particular denominations of coins. We have chosen to incorporate such attentional mechanisms in our model again because they solve several computational problems effectively, and not in order to fit psychological data. In particular, we assume that the attentional system can identify meaningful constituent parts of the scene and attend to them in turn (M.I. Posner and S.E. Peterson 1990).

A further important issue is *imagery.* In our neuroidal model the view is taken that the sensory areas of the cortex process the perceptual inputs, and transform them to more and more abstract representations as the information is passed up to higher levels. At some point nodes corresponding to "chairs" will fire. If by some internal deductions or associations the nodes corresponding to tables are caused to fire as a result, the question arises as to what then ensues that might make the system act on this new insight. The simplest view is that the system also supports reverse flows of information. The firing of the "table" neurons will cause some activity in the lower level sensory areas, similar to the activity induced by the sight of a table. This reverse activity corresponds to the act of "imagining" a table. Thus, while certain activity in the sensory area would normally cause the "table" neuron in the higher area to fire, this view suggests that exactly the reverse is also possible and typical. The existence and location of such imagery areas have been the subject of detailed investigation both by means of psychological experiments, as well as brain scans, and the results are not inconsistent with the view taken here (S.M. Kosslyn 1980, S.M. Kosslyn and O. Koenig 1992).

A further area of psychological research is on memory capacity.

Experiments have been performed to explore the limits. In one, humans were presented with 10,000 pictures over a period of five days and tested at the end of the fifth day. It was found that in the test they could distinguish pictures seen from some others not previously seen, with reasonable accuracy (L. Standing 1973). Experiments such as this, aimed at determining the quantitative limits of cognitive performance, are clearly of substantial relevance for distinguishing ultimately among detailed computational theories.

The issue of concept formation that we discussed in the previous section goes beyond simple memorization and is more difficult to investigate. Several impressive studies have been carried out with pigeons. There is substantial evidence that after seeing many examples of a concept, such as pictures that depict water in some form, the pigeons succeed in generalizing appropriately. They can classify previously unseen pictures according to whether the generalization learned holds or not (R.J. Herrnstein 1985). Indeed, it is difficult to isolate in laboratory experiments simple learning phenomena that distinguish humans from other species. The one area in which nonhumans appear to have much more difficulty is that of learning relations (R.J. Herrnstein, 1990). In Chapter 11 we find indeed that in our neuroidal system also the handling of relations introduces an extra level of difficulty.

A completely different approach to identifying the knowledge representations and learning algorithms that are used by humans is through the study of language learning in children. Do children tend to overgeneralize or undergeneralize when using a word recently learned? Many learning algorithms have a tendency to do one or the other. For example, the simplest elimination algorithm given in Chapter 9 overgeneralizes in the case of disjunctions and undergeneralizes in the case of conjunctions. Observations on humans can be used to rule out learning algorithms that are clearly inconsistent with observation. Current evidence suggests that in humans overgeneralization is more prevalent than undergeneralization (M. Bowermann 1977, Y. Levy, *et al.* 1988).

A further related issue is "short term" or "working" memory (E. Tulving 1983). Psychologists have consistently differentiated this from long term memory. In our neuroidal system we use peripherals that correspond to imagery and working memory in order to empower the NTR to perform random access tasks and to store "long term" memory.

Finally, we mention classical Pavlovian conditioning, a striking phenomenon about which a large amount of experimental data has been accumulated (I.P. Pavlov 1928, J.E. Mazur 1990). The following is a typical experiment. A subject has an air puff blown into an eye causing it to blink (or is prompted to perform some other reflex action), and at about the same time is also presented with one of a wide variety of stimuli, such as the sight of a yellow square. It is found that if this procedure is repeated enough times then the subject will become "conditioned," so that at later times the presentation of the yellow square even in the absence of the air puff will cause blinking. This can be regarded as a random access phenomenon since the range of stimuli that one can substitute for the yellow square is apparently very large. Furthermore, instead of having just one arbitrary stimulus such as the yellow square, one can have several. The conditioned response can then be made dependent on more than one such variable. It is tempting to speculate that the learning phenomena associated with Pavlovian conditioning reflect some basic learning mechanisms at the neural level. Relating them to learning algorithms is, therefore, of considerable interest (R.A. Rescorla and A.R. Wagner 1972, R.S. Sutton and A.G. Barto 1981).

The issues described so far highlight those in which there is already some existing connection between experimental findings in psychology and the neuroidal models that we develop here. This leaves two distinct classes of questions for which similar connections have yet to be made.

The first of these concerns psychological phenomena that are well supported by experiment, but which we were not forced to introduce into the current discussion of the neuroidal model. One example of a well studied area of robust phenomena, is *priming* (E. Tulving and D.L. Schacter 1990). In a typical experiment a human subject is given a list of fifty words to read. Some time later the subject is presented with some word fragments and asked to complete each one to make a word. It is found that the subject is more likely to reconstruct a word recently seen in such a list, than an equally valid alternative, even when the subject cannot consciously recollect having seen that word. In any neuroidal system that attempts to model significant areas of cognition there may be several alternative ways of incorporating priming effects. For example, recently increased weights may be allowed to attenuate with time, so that the most recent changes always have greater relative influence than earlier ones. This may be allowed only at low levels or only at certain other levels of processing. Thus various alternative models of priming can be constructed and compared by experimentation. We return to this in §12.3.

In the reverse direction, one can delineate a second class of issues, those that have received limited experimental attention to date but are suggested as relevant by the neuroidal model. An example of such a question is whether an item in memory is represented accurately in each mode separately. For example, do there exist neurons for recognizing dogs within the vision area and each of the other sensory areas separately, or are the indicators that confirm doghood in the various modalities mixed together at a level preliminary to recognition by any one? More particularly, if we present a word fragment and a picture fragment, which separately are not enough to remind a person of the object to which they both refer, can the presentation of the fragments together do so? Our ability to solve crossword puzzles suggests a positive answer to this question,[14] which in turn suggests that concepts do have such mixed mode representations. Mixed mode representations do appear to be the more economical in terms of representation and computation, and this would appear to compensate for the accompanying loss in precision. A second issue is that of hierarchical learning. There is plenty of evidence in human learning that in order to learn one thing it is often beneficial to have learned another beforehand. There is some evidence for similar phenomena in pigeons.[15] These two issues of mixed mode representations and hierarchical learning are merely examples of questions that are suggested by the neuroidal model as promising subjects for more systematic experimentation.

Chapter 5
The Neuroidal Model

5.1 Programmable Models

We do not know how difficult the task of understanding the actual circuits of the brain will turn out to be. A simple strategy of trial and error, in which one posits and tests a succession of particular algorithms, is unlikely to work if there are too many plausible algorithms to consider. It appears that in order to make headway a model that embodies some broad flexibility is needed, so that wide classes of hypotheses can be investigated together. This means that instead of considering, one by one, theories that have some specific algorithms deeply embedded in them, we should start rather with models that are *programmable*.

There is also a second, more specific argument that favors the consideration of a programmable model here. We wish to support a greater variety of tasks than previous modelers appear to have attempted within a single system. The enormous descriptional complexity of the brain may be due as much to the variety of mechanisms incorporated as to the intricacy of any one of them. In order to describe such a variety of algorithms one needs a suitably expressive language.

With this aim in mind we shall define an idealized model of a network of neurons that we shall call a *neuroidal net*. We shall use it to model the *neuroidal tabula rasa* (NTR). Each *neuroid* in the net is defined to be a *linear threshold element*, as originally formalized by W.S. McCulloch and W.H. Pitts in 1943, but is augmented by *states* and a simple *timing* mechanism. It turns out that these apparently innocuous augmentations make all the difference in rendering the model programmable. The computational power of the net depends not only on the mechanism at the nodes, but also on the network. Hence we shall consider in detail the connectivity properties of the network in conjunction with the algorithms executing on it. In such a model two opposing constraints need to be reconciled. First, the model needs to be simple enough that there is little question that real cortical neurons are at least as powerful computationally. Then any algorithm we devise for neuroids can be construed as an existence proof that the corresponding functionality can indeed be supported by cortical neurons. The second constraint on the model is that it has to capture the essence of the computational capabilities of the brain, at least for implementing random access tasks. Although neural systems in the cortex are clearly very complex, we believe that their capabilities in the realm of random access tasks can be captured by some simple model, such as ours. The direct relevance of our model to real neural computations rests, therefore, on the hypothesis that it does satisfy these two opposing constraints simultaneously.

The nodes of the neuroidal net are individual *neuroids*. In this chapter we define the functionality of these neuroids. In the chapter to follow, we shall discuss some graph-theoretic requirements on the network that are critical to the functioning of the neuroidal net.

To emphasize that we can regard a neuroidal net as something precisely defined, we shall give a somewhat formal definition of it in the style of the theory of automata (J.E. Hopcroft and J.D. Ullman 1979). A neuroidal net will be specified in five parts, in particular as the quintuple $(G, W, X, \delta, \lambda)$. Here G is the graph describing the topology of the network, W is the set of possible *weights* that the edges of the graph can have, X is the set of *modes* that a neuroid can be in at any instant, δ is the *update function for the mode*, and λ is the *update function for the weights*. We shall elaborate on each of these notions in the next section. Such a quintuple is a complete description of the net. If the *initial conditions IC* (i.e. initial weights and modes of the neuroids) and *input sequence IS* (i.e. the timing of the firing of those neuroids that are controlled directly by the peripherals) are specified, then the behavior of the net is determined. Since the model allows for randomized update functions δ and λ, if these are used then what is determined is the probabilistic behavior of the net. We note that the treatment here of individual neurons is similar in spirit to that of some earlier modelers (J.A. Feldman and D.H. Ballard 1982).

Timing is crucially important to our model. The peripherals have the power to cause various sets of neuroids in the NTR to fire simultaneously at various times. The actual choices of the sets and the times, which we call prompts, determine the input sequence IS. There is a substantial body of experimental evidence that suggests that synchronized rhythmic behavior is a pervasive characteristic of the cortex (W.J. Freeman 1975, E. Basar and T.H. Bullock 1992) and hence that the hypothesized power of causing synchronous firings, that is ascribed here to the peripherals, is not unreasonable. Between successive prompts from the peripherals our algorithms will be expected to do only a very few basic steps, perhaps typically less than ten and most often just one or two. No global synchronization mechanism is assumed here, but it is supposed that the neuroids share common notions of a time unit, and hence can keep in synchrony for such short sequences of basic steps by following their own clocks. Even if their clocks keep slightly different times this is no problem as long as they need to keep in step for only short periods. For simplicity we shall assume in the model that the neuroids have exactly identical clocks. Timing issues will be discussed in more detail in §5.3.

5.2 Neuroids

We shall now define the five components that are needed to specify any particular neuroidal net.

First, G is a *directed graph* denoted by $G = (V, E)$, where V is a finite set of N nodes labeled by distinct integers $1, 2, \cdots, N$, and E is a set of directed edges between the nodes. The edge (i, j) for $i, j \in \{1, \cdots, N\}$ is an edge directed from node i to node j.

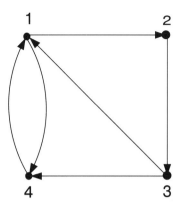

Figure 5.1. A directed graph.

We say that nodes i, j are *adjacent* or *neighbors* if at least one of the directed edges (i, j) and (j, i) belongs to E. An example with $N = 4$ and $E = \{(1,2),(2,3),(3,4)(4,1),(1,4),(3,1)\}$ is shown in Figure 5.1. A neuroid corresponds to a node j together with all the edges directed toward it (i.e. (i, j) for every i such that (i, j) is a member of E.) The intention is that an edge (i, j) models a synapse between two neurons where i is the presynaptic neuron and j is the postsynaptic neuron. By saying that a neuroid corresponds both to a node j as well as the incoming edges, we are associating each synapse with its postsynaptic neuron.

The set of edge *weights* W is a set of numbers. Each edge $(i, j) \in E$ has at each instant of time a value $w_{ij} \in W$ for its weight. Possible choices of W are the set of nonnegative integers, the set of positive real numbers, or the set of all real numbers. Alternative choices restrict the weights to some range of discrete values, such as the set $\{0, 1, 2\}$ or some range of real values such as the interval $[0, 2]$. Weights model the post-synaptic potentials, or PSPs, that were discussed in §2.3.

The *mode* of a neuroid describes every aspect of its instantaneous condition other than the weights on its edges. It models properties that are global for the neuron rather than relating only to particular synapses. It is specified as a pair $(q, \underline{T})$ of values where q is a member of Q, a finite set of *states*, and $\underline{T}$ is a vector of γ

numbers $T^{(1)} = T, T^{(2)}, \cdots, T^{(\gamma)}$ for some fixed integer γ. The set of modes, X, is the set of all such pairs. Q describes a finite choice among states. For example a neuroid that has not yet been assigned to a task or item would be typically in a different state from one that has. The algorithms described in the next five chapters require few states, sometimes only two. The first component $T^{(1)}$ of $\underline{T}$ is called T for short. In general T denotes a number that is the *threshold* of the neuroid. It models the electrical potential required to be overcome to cause the neuron to fire. For neuroid i at any instant, we denote its mode by s_i, its state by q_i and its threshold by T_i. The model allows for the possibility of $\gamma > 1$, i.e. having more than one numerical value in the mode. For example we may want a second number that expresses the confidence in a generalization that has been learned inductively by a neuroid. Where not stated otherwise we shall assume that $\gamma = 1$ and hence that s_i is the pair $[q_i, T_i]$.

The states have names that are mnemonics for their function. For example AR will be the "available relay" state. Also, Q consists of two kinds of states called *firing* and *quiescent* states. The last letter of a state name will be F if and only if it is a firing state. Also we shall define the Boolean variable f_i to have value one or zero depending on whether node i is in a firing state or not.

The *mode update function* δ and the *weight update function* λ embody the essence of the algorithm executing on the neuroid by specifying the updates that occur. Their action on neuroid i depends on, among other things, the quantity w_i. This is defined to be the sum of those weights w_{ki} of neuroid i that are on edges (k, i) that come from neuroids that are currently firing. More formally,

$$w_i = \sum_{\substack{k \text{ firing} \\ (k,i) \in E}} w_{ki}.$$

The mode update function δ defines for each combination (s_i, w_i) that holds at time t, the mode $s_i' \in X$ that neuroid i will transit to at time $t + 1$. The weight update function λ defines for each weight w_{ji} at time t the weight w_{ji}' to which it will transit at time $t + 1$, where the new weight may depend on the values of each of s_i, w_i, w_{ji}, and f_j at time t. These two transition functions can be written therefore as

$$\delta(s_i, w_i) = s_i', \text{ and}$$
$$\lambda(s_i, w_i, w_{ji}, f_j) = w_{ji}'.$$

As an example suppose that we have a neuroid i that at time $t = 0$ is in a state we call A1, and has all weights w_{ji} incoming from neighboring nodes equal to one. Suppose that, independent of the initial value of the threshold T_i, we want that at time $t = 1$ the state of i be A2, that all w_{ji} from neighbors not firing at time 0 be set to 0, and that the new threshold T_i equal the number of the neighbors j that did fire at time 0. Figure 5.2 illustrates the updates that are required in one particular instance. The transitions below show how the algorithm that realizes these updates would be expressed within the model:

$$\delta([\text{A1}, T_i], w_i) = [\text{A2}, w_i] \quad \text{for all } T_i, w_i, \text{ and}$$
$$\lambda([\text{A1}, T_i], w_i, w_{ji}, 0) = 0 \quad \text{for all } T_i, w_i, w_{ji}.$$

The first transition updates the mode. It says that if the state is A1, then the new state will be A2 and the new threshold will be w_i (which equals the number of presynaptic neighbors that were firing since at time $t = 0$ each $w_{ji} = 1$). The same update occurs for all values of T_i.

The second transition updates the value of weight w_{ji} to 0 for every w_{ji} such that the given conditions hold, namely that the state q_i is A1, and that $f_j = 0$.

As explained in later chapters, when writing more elaborate algorithms we shall, for the sake of clarity, use a more succinct notation that groups together the transitions that are to occur simultaneously at a node. This example would then be written as

$$\{q_i = \text{A1}\} \Rightarrow \{q_i := \text{A2}, T_i := w_i, \text{ if } f_j = 0 \text{ then } w_{ji} := 0\}.$$

What this algorithm achieves is that it enables node i to recognize at later times the same pattern of inputs that it was exposed to at time 0, in a certain sense. If all the inputs that were firing at time 0 fire at a later time, then the same value of w_i will be

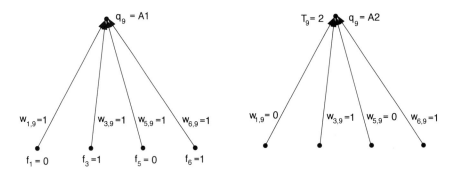

Figure 5.2. Illustration of the algorithm when the target node 9 has directed edges from nodes 1,3,5 and 6, and if initially at time 0 nodes 3 and 6 are caused to fire. The diagram on the left shows those aspects of the condition at time 0 that are relevant to the algorithm. The righthand diagram shows the result of the algorithm at time $t = 1$. At all subsequent times, whenever both of nodes 3 and 6 fire, the value of w_9 will reach the value of the threshold T_9, namely two.

achieved as at time 0, and the value of the threshold T_i set at time 1 will be reached. As we explain later, this will cause the node to fire if a standard threshold firing transition is assumed to be present also. The firing of additional inputs will not affect this outcome since the weights from these are now zero. Once the node is in state A2, no further opportunities for weight changes are possible unless further transitions are added to the specification.

Since each synapse of a real neuron appears to be either excitatory or inhibitory, we shall assume that the weights have *fixed-sign*, in the sense that each w_{ji} is predetermined to be either nonnegative, meaning that it can be never assigned a value less than zero, or nonpositive, meaning that it can never be assigned a value greater than zero.

The definitions of δ and λ express the basic intentions that (i) the updates to a neuroid should depend only on its own condition and that of neuroids from which it has incoming edges, (ii) the firing condition of these neighbors should be the only aspect of their instantaneous description that has any direct effect, and that (iii) the actual dependence on these neighbors be through a linear sum of the form in which w_i is defined above. These model in a simple way the influence of the presynaptic neurons on the postsynaptic neuron through the potentials they contribute.

Note that we allow updates to a neuroid to occur even when it is not firing itself. This is useful for preventing unwanted cascades of firings since if every neuroid that we wished to update was forced to fire, then the effects of this additional activity would also need to be controlled.

Also note that the definitions are not the minimal ones that achieve a certain expressive power. They incorporate some redundancy in order to provide for ease of expression, which is useful if the model is to be used as a programming language. For example, the dependence of λ on w_i is redundant in the sense that the value of w_i can be stored as a $T^{(k)}$ component of the mode s_i, and updated by means of δ transitions in a subsequent step. Transformations such as this, that leave the expressive power of the model invariant, serve to provide some evidence of the robustness of the model and are explored more fully in the Exercises at the end of the book. The expressive power of a robust model should be preserved under reasonably wide ranges of mutations to the model, and should not depend significantly on any arbitrary choices.

For economy of descriptions it is sometimes useful to introduce additional redundant notation. Each state $q \in Q$ has a *latency* $\ell(q)$ that is a positive integer. If node i arrives in state q at time t then neither its mode nor its weights can be changed until time $t + \ell(q)$ at the earliest. The latency is a timing mechanism corresponding roughly to the refractory period in real neurons. The updates to the mode and weights of a neuroid happen instantaneously at integral units of time. What happens at time t is determined entirely by the description of the net at time $t-1$, except that if at time t a neuroid has not completed the latency period it entered most recently, then no update will occur. Clearly, a state that has latency ℓ can be simulated by a sequence of ℓ states each of latency one, which have the property that they each go in unit time to the next state in the sequence, independent of all other conditions. Hence allowing states to have differing rather than the same latencies does not increase the power of the model, but may allow for more succinct descriptions of some algorithms. We shall assume that each state has latency one throughout this volume.

Also for economy of descriptions it is sometimes convenient to assume certain defaults. In particular, we shall assume *threshold transitions* by default. Such a transition occurs whenever all three of the following conditions hold: $w_i \geq T_i$, there is no explicitly stated other available transition, and i is not within a latency period. In such a threshold transition only the state is updated and the update changes the state to the one with the same mnemonic but with an "F" appended, indicating that it is a firing state. For example, the available relay state AR would go to state ARF. Threshold transitions typify the process by which circuits recognize inputs, while the other transitions typically modify weights or thresholds for the purpose of learning.

We shall similarly assume that a firing state has a default transition that makes a node in such a state cease firing at the end of a latency period and go into the state corresponding to its mnemonic but with the F terminator removed. Thus if node i is in state ARF at time t and if no explicitly stated transition is applicable then its state will be AR at time $t + 1$ if the latency of this state is one.

The only factors that determine the history of a net, other than its quintuple $(G, W, X, \delta, \lambda)$, are the initial conditions IC and the input sequence IS. The former specifies the mode and weights of each neuroid in the system at time $t = 0$. We envisage that a small fixed number of distinct neuroid types, each initially having some distribution of modes and weights, and occurring in some given proportions, will suffice as a typical specification of IC. For simplicity of definition we can equivalently regard all the neuroids as being identical, disjoint behavior for the different types being guaranteed by different initial states. The input sequence IS is a sequence of sets of neuroids that specifies for each $t = 0, 1, 2, \cdots$, the set of neuroids that are forced to fire (or prevented from firing) at that time by mechanisms outside the net, namely the peripherals. Such prompts or forcings comprise the only means of communicating information to the net from outside of itself.

There are two important further aspects of the model that require more involved notation in a full formal description. For simplicity we suppress this extra notation when not discussing these aspects. First, in the full model we allow the graph to be a *multigraph*. Instead of there being at most one edge from node i to node j

there may be several, say k, in which case we distinguish them by $(i,j)^1, (i,j)^2, \cdots, (i,j)^k$, and their weights by $w^1_{ij}, w^2_{ij}, \cdots, w^k_{ij}$. Such a multiplicity of edges from node i to node j corresponds to the axonal branching of neuron i having k synapses with the dendritic tree of neuron j. Note that while two synapses of weight one may have the same effect as one synapse of weight two as far as the conditions that would make j fire, the update rule λ may treat them differently. Therefore, our formal model is based on multigraphs rather than graphs. We shall, however, use the simple graph model for brevity when it makes no difference. We note that sections §14.2 and §14.3 do use the multigraph property, and that the multiple synapses that they model is crucial to the theory described there.

The second additional aspect of the model is that it allows for randomized transitions δ and λ. This means that for each combination of argument values of δ and λ there may be not just one outcome at the next time unit, but several. The neuroid will choose randomly among these according to predetermined probabilities. While randomization has found applications in many areas of computation, its role in the brain is not known. The actual algorithms we shall describe here use it in only one restricted role, that of allowing a set of neuroids to select a certain approximate fraction of themselves by having them make independent random choices.[16]

To complete the definition of the model we need to discuss what restrictions are reasonable to impose on W and on the update functions δ and λ. First we observe that, at the risk of allowing too much power, we can opt for a simple complexity theoretic model. For example, we could allow the weights w_{ji} and the thresholds T_i (or other components of $\underline{T}$) to take values from some fixed set of numbers where the size of the set grows polynomially with N, and then allow δ and λ to be *arbitrary* functions of their arguments.[17] Then the number of symbols required to describe a neuroid totally, including its transitions, would be also bounded by a polynomial in N. This is a possible choice if one favors mathematical simplicity.

The algorithms we shall describe all use much less computational power than such a definition would allow. In particular the dependence in δ of the new state q_i on w_i or $T_i^{(k)}$ is restricted to simple thresholding (i.e. applying the Boolean predicate "greater than or equal to" to it) and finite Boolean combinations of these predicates. Note that where only integer values are allowed for T, we can encode an equality condition "$T = k$" this way if by saying "$T \geq k$" and "not $T \geq k + 1$." When changes to w_{ji} update their previous values, they do so by counting up or down a sequence in some manner, e.g. $w_{ji} := w_{ji} + 1$ or $w_{ji} := 1.3 \times w_{ji}$. (More generally, for some monotone function $\triangle$, $w_{ji} := \triangle(w_{ji})$ or $\triangle^{-1}(w_{ji})$). The updates to the threshold $T_i^{(k)}$ are similarly restricted. In addition, w_{ji} or $T_i^{(k)}$ may be also assigned values that are simple arithmetic functions of w_i. Finally, we note that all the algorithms that we shall describe work even when all the weights are fixed-sign nonnegative. The computations that result under this assumption have the advantage that cascades of threshold firings will stabilize robustly. We allow for the additional possibility of negative weights because of the substantial evidence for inhibitory effects in the nervous system (I.P. Pavlov 1928). The perceptron algorithm described in §9.5 can accommodate negative weights without requiring them.

The variant of the neuroidal model that these restrictions on δ, λ and W specify is just one of several that deserve further study. Identifying sets of restrictions that capture cortical neurons more faithfully would certainly be of great interest. Presumably in biological neurons there is some bound on the number of possible states, and weight updates have some particular character such as being additive or multiplicative. Hence, in certain directions the variant, as described, will be too powerful for biological plausibility. It would be interesting to determine whether some useful extra constraints can be stated simply. Conversely, it is possible that the model misses out on some mechanisms that real neurons support and are important to the realization of random access tasks. It may be fruitful to consider choices of δ and λ that depend on different sets of arguments and have less synchronous behavior than the definitions above allow. There is some evidence that weight changes in real neurons may sometimes attenuate with time. Extensions to λ that make weights dependent on past history in this way are of interest and can be accommodated. Some authors emphasize that synapses are unreliable. This suggests other extensions in which the values of the weights behave like random variables even in the absence of any learning.

5.3 Timing

Timing plays a central role in neuroidal algorithms. In order to make programming manageable, neuroidal systems work with two very different time scales. The individual transitions δ and λ that update single neurons work on a scale of very short intervals called *microunits*. In the definitions given in the previous section, time t was measured in terms of microunits. For orchestrating the computations in the NTR, the peripherals work on a longer time scale, measured in terms of *macrounits*. In particular, we assume that the peripherals have mechanisms for global synchronization in terms of these larger units. They are able to *prompt* the NTR by simultaneously causing to fire some subset of the neuroids in the NTR that are directly controllable by the peripherals (and possibly forcing some other subset not to fire.) A cascade of computations on the microunit time scale will then ensue in the NTR, and we shall assume that this will terminate in a stable situation before a full macrounit of time has elapsed. When this macrounit has elapsed the peripherals may prompt the NTR again with the same or a different subset of neuroids.

In order to ensure that the cascade of firings in the NTR initiated by a prompt does indeed terminate in a stable situation, several alternative approaches can be taken. Since the actual choice made is of little consequence in general, we will simply presume that one such viable choice has been made. The one major issue is that the computation performed in one cascade of the NTR needs to have an interpretation that is robust to the various demands that may be made on the NTR. For example, the peripherals may prompt low level neuroids, and the algorithm being executed may need to modify neuroids representing higher level concepts that are separated in the network from the prompted ones by several intermediate neuroids. We need to avoid situations in which an algorithm is executed incorrectly because the various input signals that were to arrive simultaneously had traversed paths of different lengths, and for that reason failed to arrive when needed. In order to keep the algorithms as simple as possible we shall assume that the implementation is equivalent to one in which a neuroid undergoes cascade transitions only when all the signals that will arrive there have arrived. Clearly a sufficient condition for this is that the graph formed by the neuroids that are actively involved in any cascade is acyclic, and all paths from the inputs to any one node are of the same length. However, the latter condition, on path length, becomes redundant if each signal is a long enough train of spikes that every neuroid in the cascade has time to reach some stable level of activity, and if it is assumed in addition that in any such cascade no negative weighted edge is involved that might have the effect of reducing firing activity.

A *neuroidal algorithm* for a task, such as unsupervised memorization, will be defined as a sequence of *steps* each initiated by a prompt from the peripherals and resulting in some updates to some neuroids. These prompts will be separated in time by one, or a larger whole number of macrounits. It turns out that neuroidal algorithms of the kind in which we are interested have convenient *high-level descriptions* which describe their essence. In these the possibility that the prompted neuroids influence the ones taking part in the algorithm through a cascade of threshold firings, is suppressed. It is assumed that any items that need to be prompted by the peripherals can be prompted directly by the peripherals. The result is that a macrounit can be equated with a microunit in the description of the algorithms. Hence we shall describe all algorithms at this level on just one time scale. A more detailed level of neuroidal implementation, of the kind described in the previous paragraph, is assumed to support these high level algorithms. Perhaps the simplest way to interpret our algorithms, therefore, is to assume that the threshold transitions that occur in cascades take infinitesimal or zero time, while all other transitions take unit time, corresponding to a macrounit. In order to be consistent with this, in the high level descriptions we shall use the convention that threshold transitions take zero time.

To summarize, *all transitions in the algorithms to be described will be assumed to take unit time, with the exception of the threshold transitions and prompts, which will both cause the corresponding nodes to fire instantaneously.*

A stylistic point is that it may make it easier to implement high level algorithms if information is not encoded in the firing status of neuroids between steps. Then the only neuroids that fire at the start of a step are those that are prompted by the peripherals, and those that undergo threshold firing as a result.

An *interaction* is the computation resulting in the NTR from an input from the outside world, or from certain peripherals that model such things as imagery or short term memory. In either case the input is processed (by some peripheral) and the information so derived presented to the NTR in accordance with some neuroidal algorithm that consists of a number of *steps* as described above.

Once G, X, and W are defined for the NTR, an algorithm for a functionality is a specification of the update functions of the neuroids as well as a description of the input sequence that has to be applied by the peripherals, that together enable the functionality to be realized. Such an algorithm may require several steps. Some algorithms, particularly those for inductive learning, achieve their desired result after a sequence of interactions, corresponding possibly to a long series of encounters with the outside world.

Chapter 6
Knowledge Representations

6.1 Positive Knowledge Representations

If a neural system can cope in a complex external world, one must presume that its behavior can be described in terms of the various semantic items that are meaningful in that world. By the term *item* we mean just about any aspect of the world that may be useful in describing it. Individual objects, events, properties, relationships, concepts, and categories are all examples of items, as are certain predicates that are detected at lower levels in our perceptual systems.

It seems plausible to hypothesize that the items that are appropriate for describing the world of experience of a system provide the right vocabulary for describing its behavior when it interacts with that world. What is more controversial and problematic is to determine whether these same items also provide the most appropriate internal vocabulary for the neural system. Does the firing of a neuron in the cortex signal the recognition of one such item, or does it correspond to something that is best described in completely different terms? Conversely, does each item in the world that can be recognized by an individual correspond to some neurons that recognize it?

This central issue has received much attention. Barlow gives an early comprehensive treatment of the "localist" position that takes a positive view on these questions (H.B. Barlow 1972). At the other extreme are theories that posit global, sometimes called holographic, representations (J.J. Hopfield 1982, J.L. McClelland and D.E. Rumelhart 1986), in which the representation of any item is best viewed as being spread over many or all the neurons. Various intermediate positions may also be taken (J.A. Feldman 1990).

The representations that emerge from our various analyses here are all of one flavor, and we shall call these *positive* representations. This flavor can be characterized in terms of five features that we shall label (a),···,(e), respectively.

The first two features are characteristic of the localist position and state that (a) each neuron corresponds to a semantic item, and (b) there are typically several neurons representing each such item that is represented at all. Localist positions are sometimes called "grandmother cell" theories because they assert that there are some neurons that fire if and only if one thinks of one's grandmother. The objection that is often raised to such theories is that there just are not enough neurons to represent every combination of attributes that one may need to represent. Do we have neurons to represent our grandmothers in every possible location? Using a different metaphor, do we have neurons to represent "a yellow Volkswagen" and every other combination of car makes and colors?[18]

The representation we use answers this objection by having a third characteristic: (c) only those new items are added to memory that are experienced and noticed by the attentional mechanisms. Thus on seeing a scene various circuits may fire as we attend to different parts of it and to different attributes of the parts. Unless yellow Volkswagens have figured importantly in our previous lives, neurons for both yellow and Volkswagen will fire, but there may not be neurons previously allocated for the combination. In this sense our representation has some of the characteristics of *population coding*, namely that inputs cause subsets of neurons to fire that may overlap for different inputs. In our representations two such subsets will overlap only if the corresponding inputs share some common characteristics.

A fourth aspect of the representation is that (d) it is *hierarchical*. Some items will be represented at the start and may be viewed as having been preprogrammed. Once some items have been assigned to neurons, new items expressible in terms of the items already represented can be assigned to previously unused neurons. By the word hierarchical we do not imply strict hierarchies in the sense that cyclical relationships are forbidden. It is possible that once two related items have been assigned, their meaning is refined in terms of each other. The emphasis is more on the idea that the representation is structured so that some items are high level, being satisfied for very specific inputs, while others are lower level and of greater generality, being satisfied for wider ranges of inputs. Some of the lowest level items may be determined by the basic units of the peripherals that perform perception. High level items are typically represented in terms of combinations of lower level items.

Lastly, we mention that while the intention of the representation as described so far is to express Boolean functions, the reality of the neuroidal implementations makes it *graded*, so that (e) only approximations of any idealized Boolean functions are represented. One reason for this is that the physical neural connections necessary to realize the idealized function will be present only with high probability, and not with certainty. A second reason is that there may not be any simple idealized function to realize. For example, a circuit for recognizing a chair may involve subcircuits for various particular chairs and various types of chairs so that the overall circuit has no simple characterization.

All the algorithms described in this volume work with knowledge representations of this flavor. The items represented will be denoted by letters such as x, y, z. The sets of neuroids representing each of these items will be $\tilde{x}, \tilde{y}, \tilde{z}$, respectively. The basic intention is that when the system is presented from the outside with an input corresponding to item x, then the neuroids in the set $\tilde{x}$ will fire. Typically there will be about r neuroids for each item, r being called the *replication factor*. Replication will ensure robustness, as well as a slightly lower connectivity requirement on the network than would be needed otherwise. Several authors have previously used sets or assemblies of cells to represent a concept (D.O. Hebb 1949, V. Braitenberg 1978), usually with the implication that within each such set or assembly, each member has an excitatory influence on the others. Such assemblies would have a self-imposed tendency toward an all-or-none firing behavior. In our representation we make no such assumption.

To illustrate one specific way in which the representation generated by our algorithms is graded, suppose that $x_1, x_2,$ and x_3 are already represented by nodes and we wish to represent the conjunction $y = x_1 \wedge x_2 \wedge x_3$. Now y will be represented by r nodes. Most of these will be connected via some appropriate circuitry to members of all three of $\tilde{x}_1, \tilde{x}_2,$ and $\tilde{x}_3$ and will be able to represent $x_1 \wedge x_2 \wedge x_3$. Some may be connected to only $\tilde{x}_1$ and $\tilde{x}_3$, say, and will represent $x_1 \wedge x_3$. The different members of $\tilde{y}$ may, therefore, represent the desired function with varying accuracy. Thus some grading is introduced even at the neuroidal level. If we consider how a neuroidal algorithm might be implemented by biological neurons, there are many additional potential sources of grading. For example, it may be that a spike train is interpreted more accurately as a numerical value, corresponding to some measure of frequency or duration rather than a Boolean truth value. In that case the operations on them are only approximated by the Boolean operations that we consider here.

We envisage that, in practice, any grading in the individual circuits is compensated for by having *multiple representations* for important items, so as to ensure that the overall system is highly reliable. By this we mean that the nodes corresponding, for example, to the item chair are associated with several circuits, each of which attempts to find confirmation of chairhood in a different way. If a reasonable fraction of these succeed, then it can be assumed that overwhelming evidence of chairhood has been found.

6.2 Vicinal Algorithms

All the algorithms that we shall describe can be considered, at a suitable level of abstraction, as *vicinal* or neighborly. The most basic feature of a vicinal algorithm is that, whenever some communication has to be established between two items not directly connected, the algorithm establishes the necessary communication via neuroids that are each common neighbors of some pair of neuroids that represent the two items respectively. Thus, if $\tilde{x}$ and $\tilde{y}$ correspond to the two items and $\hat{E}(\tilde{x}), \hat{E}(\tilde{y})$ are the neighboring nodes of the sets $\tilde{x}$ and $\tilde{y}$ respectively, then $\hat{E}(\tilde{x}) \cap \hat{E}(\tilde{y})$, the *undirected frontier* of $\tilde{x}$ and $\tilde{y}$, will be the set through which communication takes place. Because of the primacy they give to communication via common neighbors we call these algorithms vicinal. This approach contrasts with the various communication schemes currently proposed for parallel computers, which will typically route packets of information via a succession of intermediate nodes to a distant node with a specified identifier or address.

Some algorithms may require that a two-way channel of communication be established, if temporarily, between pairs of nodes. This is done most easily by requiring that the edges be *bidirectional*, in the sense that a directed edge $(i, j) \in E$ has an associated reverse edge (j, i) also belonging to E. We shall sometimes refer to such a pair of edges between two nodes in opposite directions as a *bidirected edge*. The algorithms that exploit bidirectionality can be interpreted as demonstrating the computational efficacy of point-to-point feedback. Whether such bidirectionality between individual neurons is pervasive in the cortex is currently unknown. However, it is well established that for most pathways linking one cortical area to another there are reciprocal pathways going in the opposite direction. What is unknown is whether the reciprocity is precise enough to realize bidirected edges directly.

In this chapter we shall describe perhaps the simplest model, namely *random graphs*, that support vicinal algorithms directly. These graphs have two important properties that are necessary for supporting these algorithms, namely a certain *frontier* property, and a certain *hashing* property. An alternative graph model, namely *random multipartite graphs* have very similar frontier and hashing properties, and are equally good for supporting vicinal algorithms. Where we assume the basic random graph case, we do so only for the sake of greater simplicity. The main difference between the models is that while the first treats all the nodes as equal, the second splits them into sets, each of which corresponds to a different area of the cortex. In both cases the edges model long distance communication. In the multipartite case we assume that certain pairs of areas are connected. Those pairs that are have random connec-

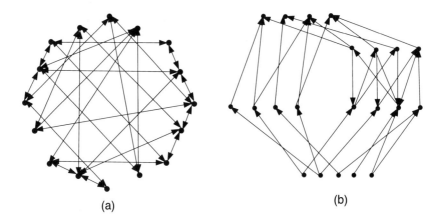

(a)

(b)

Figure 6.1. Schematic diagram of (a) a random graph; and (b) a random multipartite graph that models the connections among five cortical areas.

tions between them. Note that random multipartite graphs resemble some of the hierarchical structures that have been found, for example, in visual cortex (D.J. Felleman and D.C. van Essen 1991).

We can consider each of the two classes of graphs in each of the two cases of being directed or bidirected. Figure 6.1(a) illustrates the bidirected case for general random graphs. This is the one on which vicinal algorithms can be implemented most directly. On the other hand, vicinal algorithms can be supported also without this bidirectional assumption. Figure 6.1(b) shows an alternative such model for a random multipartite graph. The edges between two areas either all go in one direction, or, as illustrated for one pair of the areas, they go in both directions. Note, however, that in the latter case no individual edge is bidirected (except possibly a very few by chance.) We shall show in later chapters that vicinal algorithms can also be implemented on other classes of graphs. Hence these algorithms may be viewed as high level conceptual abstractions for programming the neuroidal model.

6.3 Frontier Properties and Storing New Items

The most basic aspect of our knowledge representation is that each item is stored in about r neuroids, where the replication factor r is viewed as a constant (say 50 or 100). Since vicinal algorithms establish communication between the sets representing two items through the frontier of these two sets, this frontier should be nonempty in general. Furthermore, since it is the aim of some of the algorithms to store a new item at this frontier, so that the new item becomes an equal citizen with the others, one needs that this frontier be of size about r also. For example, if $\tilde{x}$ and $\tilde{y}$ store "yellow" and "Volkswagen" respectively, then we may allocate their frontier to store "yellow Volkswagen." To be more precise we define $E(\tilde{x}), E(\tilde{y})$ to be the sets of nodes to which edges are directed from $\tilde{x}, \tilde{y}$ respectively, and call these the *directed neighbors*. We then define $E(\tilde{x}) \cap E(\tilde{y})$ to be the *directed frontier* of $\tilde{x}$ and $\tilde{y}$. As illustrated in Figure 6.2 the algorithm will allocate this directed frontier, which we will simply call the *frontier* from now on, for storing $x \wedge y$. The task of allocating nodes for storing a conjunction $x \wedge y$ is called JOIN in Chapter 14, and discussed there in more generality. The method described below and in the next chapter can be viewed as the simplest vicinal implementation of this more general task.

These considerations suggest the following definition. We say that a graph $G = (V, E)$ has the (r, l, m)-*frontier* property if, when $\tilde{x}, \tilde{y} \subseteq V$ are randomly chosen disjoint subsets of size r, the size of the frontier $E(\tilde{x}) \cap E(\tilde{y})$ has expectation l and variance m. A graph that is ideal for executing vicinal algorithms would have $l = r$ and $m = 0$.

Consider a random graph on N nodes such that for each pair (i, j) of nodes, a directed edge joining i to j is present with probability p, independent of all other pairs. Then the expected number of edges directed toward any one node is the same as the expected number directed away and equals pN. These quantities are called the expected indegree and outdegree, and where they are equal we shall call them simply the *degree*.

$$E(\tilde{x}) \cap E(\tilde{y})$$

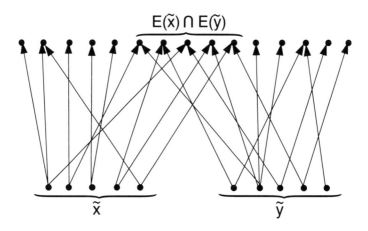

Figure 6.2. Sets $\tilde{x}$ and $\tilde{y}$ are illustrated by the leftmost five and the rightmost five nodes, respectively, in the bottom row. Their frontier, consisting of the five nodes in the middle in the top row, is where the conjunction $x \wedge y$ will be stored.

We regard r as a fixed constant, N as varying and large, and p as diminishing as N grows. Let $\tilde{x}, \tilde{y}$ be disjoint sets of r nodes. Then the probability that any one fixed node i, not belonging to $\tilde{x}$ or $\tilde{y}$, lies in the frontier of $\tilde{x}$ and $\tilde{y}$ is $(1-(1-p)^r)^2$. To see this first note that the probability that $i \notin E(\tilde{x})$ is $(1 - p)^r$, since $(1 - p)$ is the probability that it fails to be connected to any fixed member of $\tilde{x}$, and there are r such members. The probability that it is connected to at least one node in $\tilde{x}$ is, therefore, $(1 - (1 - p)^r)$. Since the same statement holds for $\tilde{y}$, the square of this expression gives the desired probability. Note that we are assuming here that the edges coming to i from different nodes have probabilities of being present independent of each other. Also, we are using the fact that the probability of several independent events occurring together is the product of the probabilities of their occurrence separately.

Here, as well as throughout this chapter and Chapter 14, we shall make the following assumptions in all our estimations of probabilities, which we call the *pristine conditions assumptions*. First, the number of neuroids representing any one item already stored (i.e. the size of $\tilde{x}$ or of $\tilde{y}$ in this case) is exactly r, and, second, that the edges directed toward neuroids not yet allocated are present with equal probability independently. Once a chain of allocations has been made by our methods, these assumptions will hold only approximately.

Regarding r as fixed,[19] if $p \to 0$ as $N \to \infty$, then $(1-(1-p)^r))^2$ differs from $r^2 p^2$ by a quantity that diminishes proportionately with p^3. In more convenient notation one would say that $(1-(1-p)^r)^2 = r^2 p^2 + O(p^3)$. This follows by application of the Binomial Theorem to $(1-(1-p)^r)^2 = 1 - 2(1-p)^r + (1-p)^{2r} = 1 - 2(1 - rp + r(r-1)p^2/2 + O(p^3)) + (1 - 2rp + 2r(2r-1)p^2/2 + O(p^3)) = r^2 p^2 + O(p^3)$.

Then for each i not in $\tilde{x}$ or $\tilde{y}$ the probability of being in their frontier is given by $p^* = r^2 p^2 + O(p^3)$ and is independent for different choices of i. Hence the size of the frontier is governed by a binomial distribution consisting of $N^* = N - 2r$ trials each with probability p^* of success. If we choose $p = (Nr)^{-1/2}$ then we obtain that the number of nodes in the frontier has

$$\text{expectation} = N^* p^* = (N - 2r)(r/N - O(N^{-3/2}))$$
$$= r - O(N^{-1/2}), \text{and}$$
$$\text{variance} = N^* p^* (1 - p^*)$$
$$= (N - 2r)(r/N - O(N^{-3/2}))(1 - r/N + O(N^{-3/2}))$$
$$= r - O(N^{-1/2}).$$

We conclude that a random graph with $p = (Nr)^{-1/2}$, or expected degree $(N/r)^{1/2}$, has the $(r, r - O(N^{-1/2}), r - O(N^{-1/2}))$-frontier property. For implementing vicinal algorithms random graphs of this degree appear therefore to be ideal for ensuring that the expected size of the frontier of two sets $\tilde{x}, \tilde{y}$ of size r will be about the same as their own size and, therefore, that this frontier has some chance of storing a new item that is to become an equal citizen with the items x and y.

Since we wish that the NTR be able to learn hierarchically, we have to ask what happens if we create new items out of such frontiers and repeat this process, creating new frontiers from sets that were previously frontiers themselves, to arbitrary depth. It turns out that the variance of this process is too large to maintain stability over a large number of iterations. Experiments suggest, however, that for $r = 50$, for example, this process is maintainable in such random graphs to depth 4 or 5 (A. Gerbessiotis 1993). It is an open question whether there exist graphs that have frontier properties that have lower variance and hence are more stable. It is known that for $r \geq 2$ the ideally stable $(r, r, 0)$-frontier property is not possessed by any bidirected graph with more than $3r$ nodes (A. Gerbessiotis 1993).

There are several plausible approaches to hierarchical learning that sidestep this instability problem. One simple approach is to replace the random graph model by a random multipartite model of fixed depth, as illustrated in Figure 6.1(b). As already noted, if the connections between successive layers are random then most of the beneficial properties that random graphs provide are retained, except now, we have the added benefit that the depths of all the hierarchies are limited to the number of layers in the graph.

A more interesting twist comes from observing that it is only in the process of allocating neuroids to new items that the hierarchy depth needs to be restricted. Once these allocations have been made, associations can be established in supervised mode (as in Chapters 8 or 9), with arbitrary implications for the depths of the semantic hierarchies. This suggests that before an item is stored it should be given a *name* from some simple uniform space of names, such as a sequence of phonemes or letters. Such a space of names should be simple enough that there be a mechanism for allocating arbitrary names to neuroids in a multipartite network of small depth. Then when we learn a new concept, such as that of a "dinosaur", we allocate nodes to it according to a shallow hierarchy, that is sufficient since the structure of the sequence of phonemes or letters in such a word is necessarily limited (see Figure 12.1 in Chapter 12). We can then add an arbitrarily deep hierarchical definition of the semantics of this item by learning in supervised mode relationships between the "dinosaur" item and some of the other items that are also represented. According to this approach language, and in particular, naming, acquires a special role in cognition. This role can be summarized as follows: Hierarchical learning may not be feasible if storage for items is allocated solely according to their semantics, since that may give rise to hierarchies that are too deep to be supported stably by the network. Hierarchical learning may become feasible, however, if the items are encoded in some more uniform manner, as by sound pattern, for which a fixed depth mechanism for storage allocation is sufficient.

6.4   Frontier Properties and Associations

The common neighbors of pairs of neuroids play another important role in vicinal algorithms besides memory allocation. This second application is to establishing *associations* between arbitrary pairs of items already stored. If node sets $\tilde{x}$ and $\tilde{z}$ are already allocated, we may wish to update the network so that at later times whenever $\tilde{x}$ fires so will $\tilde{z}$. In Chapter 14 we will call this update operation LINK. To achieve it in the simplest way we can attempt to assign to a set $\tilde{y}$ the role of *relay* nodes, and change some weights incoming to $\tilde{y}$ and $\tilde{z}$ in such a way that at a later time the firing of $\tilde{x}$ will influence the firing of $\tilde{z}$. This is illustrated in Figure 6.3. The algorithms of Chapters 8 and 9 for supervised learning are all based on some such implementation of this LINK operation. The question we ask is the following. Suppose that $\mu \geq 0$ is some constant and r and N are as before. If the graph is random with edge probability $p = (\mu/(rN))^{1/2}$, what is the expected number of members of $\tilde{z}$ that have a common neighbor with at least one member of $\tilde{x}$? As we shall see, if $\tilde{x}$ and $\tilde{z}$ have size r, then the answer is about $r(1-e^{-\mu})$. For example, if $\mu = 1$, then about 63% of the $\tilde{z}$ will be connected to some member of $\tilde{x}$ via a common neighbor, while if $\mu = 2$ (which increases the degree by a factor of 1.414), then the proportion is as much as about 86%. If $\mu = 4$ (i.e. degree is $2(N/r)^{1/2}$), then the fraction is larger than 98%. We note that by having values for μ greater than one we perturb the frontier node allocation process described in Section 6.3, in that we will be allocating about μr rather than r nodes to each new item. We can counteract this by, for example, having the node allocation process reject each node provisionally allocated with probability μ^{-1} randomly. In this way we can support on the same network both the allocation process JOIN as well as the process LINK that establishes associations.

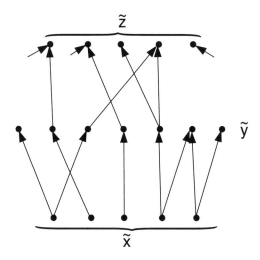

Figure 6.3. Illustration of an association set up from node set $\tilde{x}$ to node set $\tilde{z}$ via a set of relay nodes $\tilde{y}$. In the illustration there are four $\tilde{y}$ nodes that serve as relay nodes. Also, the association is made successfully with four of the five members of $\tilde{z}$. The remaining member lacks the necessary connections.

To prove our claim about the number of members of $\tilde{z}$ that have a common neighbor with $\tilde{x}$, we first consider the probability that there is a path of length two to one fixed member, say $i \in \tilde{z}$, from at least one member of $\tilde{x}$ via a fixed node j. Since the probability that at least one member of $\tilde{x}$ is connected to j is $(1 - (1 - p)^r)$, as above, and the probability that j is connected to i is p, the probability of both happening is $p(1 - (1 - p)^r)$. Using the Binomial Theorem, this can be written as $p(rp + O(p^2))$, which, after substituting for p gives $\mu/N + O(N^{-3/2})$. Hence the probability that there is such a path via at least one of the $N - 2r$ possible choices of j is

$$1 - (1 - \mu/N + O(N^{-3/2}))^{N-2r}.$$

Using the fact that $(1 - \frac{1}{x})^x \to e^{-1}$ as $x \to \infty$ we obtain[20] that this probability is $1 - e^{-\mu}(1 + O(N^{-1/2}))$. Since this probability holds for each $i \in \tilde{z}$, the expected number of members of $\tilde{z}$ having the required property is r times this quantity, which approaches $r(1 - e^{-\mu})$ asymptotically as $N \to \infty$.

The conclusion of the analysis of this and the previous sections, for random directed graphs with $p = (\mu/rN))^{1/2}$ and $\mu = 4$ is the following. When allocating storage to a new item (by an algorithm such as Algorithm 7.2) one half of the frontier set will be retained for the purpose of storing the item and the rest will be freed. When establishing an association from the representatives of one item $\tilde{x}$ to the representatives of another $\tilde{z}$ (by an algorithm such as Algorithm 8.1 or 8.2), an expected fraction of $(1 - e^{-4}) \approx 0.98$ of the nodes of $\tilde{z}$ will be reached successfully.

6.5 Hashing

Besides the frontier properties, there is a further attribute that is required of the network and can be supplied by randomness. This further property is needed to ensure that the nodes chosen for representing a new item will be, to large measure, among those not previously chosen. We shall call this the *hashing* property since it corresponds to that notion in computer science.

The only mechanism we use for allocating new storage is the one described earlier, that of assigning a frontier $E(\tilde{x}) \cap E(\tilde{y})$ to a new item that is associated with the conjunction $x \wedge y$. The property that needs to be ensured is that for any choice of x and y, this frontier contains a sufficient number of previously unallocated neuroids that these are able to represent the new item effectively. As in conventional hashing we shall assume that only a certain constant fraction of all the neuroids are ever assigned. The allocation mechanism will degrade severely as the memory fills up beyond a certain fraction. For plausibility we observe that a biological system, such as a human, living for 100 years and having 10^{10} available neurons will be able to allocate up to 10^3 new neurons each hour without more than 10% of the memory ever filling up. This holds even in the absence of any provisions for freeing memory or forgetting.

To conceptualize the hashing process we take the view of universal hashing[21] where the sequence of data requests over the lifetime of the system is taken as fixed, and the hash function, or in our case the graph, is taken as chosen randomly and independently of the requests. This viewpoint allows us to analyze the behavior for arbitrary sequences of memorization requests without any seriously limiting assumptions having to be made about the nature of the sequence. In particular, we shall consider the situation in which a set of items is already stored and a new one is to be allocated. We allocate the frontier of some appropriate pair of sets $\tilde{x}$ and $\tilde{y}$ to store the new item. This pair is arbitrary (though perhaps restricted to a certain depth in the network as previously mentioned). We need to show that if the graph was randomly chosen then with high likelihood it can accommodate this new item in the sense that the frontier will contain a significant number of nodes that are still free and available for allocation. The basic phenomenon is the same as that exhibited in Section 6.2, except that now only a certain fraction of the nodes are still available.

We can do a very approximate calculation using the pristine conditions assumptions of Section 6.3 and assuming $\mu = 1$ for simplicity. We consider $\tilde{x}$ and $\tilde{y}$ both to have size r, assume that there are $N(1 - u)$ nodes still unallocated for some constant u ($0 \leq u \leq 1$), and assume that each potential edge to any fixed unallocated node i, from any fixed $\tilde{x}$ or $\tilde{y}$ node, is present with probability $p = (Nr)^{-1/2}$ independent of the presence of other edges. Then, exactly as before, the probability that node i is adjacent to some node in $\tilde{x}$ and some node in $\tilde{y}$ is $(1-(1-p)^r)^2 = r^2p^2+O(p^3)$. Hence, if there are $N(1 - u)$ choices of i then the expected size of the frontier of $\tilde{x}$ and $\tilde{y}$ that is unallocated is

$$
\begin{aligned}
& N(1 - u)(r^2p^2 + O(p^3)) \\
= \ & N(1 - u)\left(r^2\frac{1}{Nr} + O\left(\frac{1}{(Nr)^{3/2}}\right)\right) \\
= \ & r(1 - u) + O(N^{-1/2}).
\end{aligned}
$$

What such a hashed memory achieves can be viewed as follows: Regarding r as a constant, suppose that we have $(u/2)N/r$ items already stored, for some small enough constant $u < 1$, and some sequence of another $(u/2)N/r$ new items come along, each one expressible as the conjunction of a pair previously memorized. Although the new pairs may be viewed as chosen from many more (proportional to N^2) pairs potentially, and although only N neuroids are available, we can nevertheless allocate memory for this arbitrary sequence of new items that present themselves, at least with high probability. The main point is that we can allocate these arbitrary new items drawn from quadratically many possibilities, without needing quadratically many nodes!

In order to capture the precise behavior of this hashing mechanism we need to be more careful in the analysis than the brief argument given above. There is clearly a contradiction in making the pristine conditions assumption here, since in discussing hashing we want to make a claim about what happens as the memory fills up. In particular, the items we can accommodate in this manner are clearly constrained in the following way: No item x can occur in more than about $(N/r)^{1/2}$ distinct new items, since there are not enough neuroids adjacent to $\tilde{x}$ to represent them all. It is, however, the case that for some constants $0 < u, v < 1$ if any uN/r conjunctions need to be stored and no item occurs in more than $v(N/r)^{1/2}$ of them, then they can all be allocated with high probability, assuming only a certain hierarchical stability that guarantees each allocated item has close to r nodes. The aspect of the pristine conditions assumption that requires that edges directed toward unallocated nodes have the same probability of being present independently of each other is not required.

The following is a brief outline of the idea needed to verify this. As we said previously, in general we view the sequence of requests as fixed and consider the graph as being generated randomly independent of the requests. We can do this generation more carefully by considering each allocation request in turn, and at each stage generating only those edges of the graph that are strictly necessary. Thus if the conjunction of x and y is required then we will randomly choose the nodes that will represent $x \wedge y$ according to the correct probability. The end points of all other edges going from $\tilde{x}$ or $\tilde{y}$ to unallocated nodes will remain uncommitted and unbiased (although the probability of any one of these edges existing will then be slightly biased to below p).

We then consider a stage in the above process when fewer than

$uN/2$ nodes have been allocated and hence at least $(1 - u/2)N$ remain free, where N was their total number initially. For each allocated node, among the $(N/r)^{1/2}$ edges directed away from it at most a fraction v were responsible for actual allocations, and at most another (expected) fraction u go to allocated nodes for whose allocations they were not responsible. Hence the expected number of available nodes allocated to $x \wedge y$ can be computed by exactly the same method as in §6.3. This will give some constant fraction of r rather than r, since the number of available neuroids is now only a constant fraction of N, and the edge probability is biased away from p by a constant fraction depending on u and v.

Chapter 7
Unsupervised Memorization

7.1 An Algorithm

Unsupervised memorization, as formulated earlier, appears on the surface to be a simple task. Following one presentation of an input, changes in the neuroidal system take place so that if an identical or similar enough input is presented in the future, the neuroidal system will recognize this repetition. By the term similar, we are not implying any abstract metric of similarity. We mean simply that the effects on the other circuits present in the system are substantially the same.

Despite its apparent simplicity this task poses a very fundamental problem, that of storage allocation. Since the instance to be memorized may be unanticipated and, within limits, arbitrary, a mechanism is needed for allocating storage space to essentially arbitrary new items. Since we employ a positive knowledge representation we treat the process of storage allocation as one of identifying some previously unused neuroids and committing them to the purpose of representing this new item. Such a strategy, if it can be implemented, makes it possible for learning to be *cumulative*, in the sense that unrelated items in memory will be left alone by any one execution of unsupervised memorization. The relative ease of dealing with this issue in positive representations is an important factor in our favoring it over the alternatives.

As discussed earlier we regard memorization as being related to forming Boolean conjunctions. We shall, therefore, consider the following idealization of the task at hand. An input has a number of attributes that correspond to items $x_1, \cdots, x_n$ that are represented in the NTR by neuroid sets $\tilde{x}_1, \cdots, \tilde{x}_n$. When this input is presented the nodes in all the sets $\tilde{x}_1, \cdots, \tilde{x}_n$ will be caused to fire. The problem, in its purest form, is to allocate a new set $\tilde{z}$ of neuroids to represent the new item z that corresponds to this input, and to update the net in such a way that in future interactions the nodes $\tilde{z}$ will fire if and only if $\tilde{x}_1, \cdots, \tilde{x}_n$ all fire simultaneously. We regard these sets as all having size r, and regard the neuroids within any one of these sets as having identical behavior (i.e. at any time either all fire or none do). In later chapters we shall discuss the more realistic and robust situation in which the item represented by such a set is considered as recognized if the fraction of its elements that are firing is above a certain value. The issue of how the peripherals may mediate and influence the choice of attributes that are presented to the NTR will also be discussed in later chapters.

We note that here, as elsewhere, all items once allocated are regarded as equal citizens. Items represented by nodes that are allocated by the mechanisms of this chapter play the same role as items represented by nodes that are controlled directly by the peripherals. Except for the proviso that some limits on the depth of the hierarchies of allocations may be imposed, the theory defines items *relative* to each other rather than absolutely. To some extent the hierarchical memory allocation process implies a hierarchical view of the knowledge that can be stored. As noted previously, however, any item already stored can be learned in supervised mode in terms of any of the others. Hence circular relationships among the items may develop that do not respect the hierarchy of the original allocation process.

We shall first discuss the case of $n = 2$, which we call 2-conjunctions. The more general case can be reduced to this one. The solution we propose is the following: We assume that the network has the frontier properties described in the previous chapter. For learning $z = x_1 \wedge x_2$ we shall simply allocate to $\tilde{z}$ the frontier $E(\tilde{x}_1) \cap E(\tilde{x}_2)$, or rather those members of this frontier that are in a certain state which we call "available memory" state, indicating that they are not already allocated. What remains then is to specify an algorithm that will enable the neuroids in this $\tilde{z}$ to modify themselves so that at later times they will fire whenever both $\tilde{x}_1$ and $\tilde{x}_2$ fire simultaneously.

The neuroids that implement unsupervised memorization are all initially in *available memory* state (AM). They are in readiness to be called upon to store an item. Their initialized condition is that their edges all have weight one and their threshold is effectively infinite, or, in other words, higher than the number of inputs. They cannot undergo threshold firing until after they are allocated and have changed to *unsupervised memory* state (UM).

First we describe an initial attempt at an algorithm. This will serve to further illustrate the notation we use for algorithms in general, as well as to explicate the correct version that follows. In this first version the algorithm is prompted by the peripherals firing a set $\tilde{I}$ of neuroids. Any neuroid in state AM, say neuroid i, that is connected to at least two members of $\tilde{I}$ will change to state UM. It will undergo threshold firings at later times whenever the nodes in $\tilde{I}$ to which it is connected all fire simultaneously.

Algorithm 7.1

Step 0: Prompt : $\tilde{I}$.

$$\{q_i = \text{AM}, w_i \geq 2\} \Rightarrow$$

$$\{q_i := \text{UM}, T_i := w_i, \text{if } f_i = 0 \text{ then } w_{ij} := 0\}.$$

In general we describe algorithms in the format of a sequence of steps. At the tth step "Prompt" describes the set of neuroids that the peripherals force to fire (or prevent from firing) at time t macrounits after the start. The transitions that are described following this are those that are expected to be invoked during the time unit that follows at the relevant neuroids (unless prevented by latency). In order to effect the prompt itself some global orchestration will be required from the peripherals. All other aspects of the algorithm are fully distributed. Its course at any neuroid not directly prompted is completely determined by the transition rules and by the conditions at that neuroid and at its immediate neighbors. The overall algorithm can be invoked at any time by having the preconditions of the first step satisfied as a result of an appropriate prompt.

As explained earlier, for economy of notation we shall describe as a single rule the transitions that can be invoked by a neuroid for any one precondition. The lefthand side describes the necessary precondition. The righthand side describes the update to the mode, as well as the updates to all the weights. The latter may take the form of a conditional statement since the update to each weight may depend on whether the corresponding adjacent neuroid is firing. In this way we shall write on a single line a combination of δ and λ transitions. *The conditions in a rule always refer to time t, while the changes produced happen at time $t + 1$.* Default threshold transitions are instantaneous but are implied to exist *only* for modes that do not appear in a precondition in any explicitly stated rule.

Algorithm 7.1 will provide the following behavior to any neuroid i that is initially in state AM and has all incoming weights equal to one. The neuroid will be inactive until the first occasion when at least two of its presynaptic neighbors fire. When that happens $w_i \geq 2$ will hold, since each incoming weight is one and their sum over all edges coming from firing neighbors will then be at least two. If this condition is first satisfied for the neuroid at time t, then three changes to the neuroid will occur simultaneously one unit of time later. Its state will become UM, its threshold will become equal to w_i, and the weights on the edges that come from nodes not firing at time t will be made zero. The first of these is intended to ensure that no attempt to allocate this neuroid at any later time will succeed. The last two changes ensure that the neuroid will undergo threshold firing whenever in the future its neighbors in $\tilde{I}$ that fired at time t, fire again.

This algorithm can be seen as a first attempt at learning the conjunction $x_1 \wedge x_2$, if we interpret $\tilde{I}$ as $\tilde{x}_1 \cup \tilde{x}_2$. For nodes connected to at least one member of $\tilde{x}_1$ and at least one member of $\tilde{x}_2$, the algorithm will work as required, as illustrated in Figure 7.1. Unfortunately, nodes connected to two nodes in $\tilde{x}_1$ and none in $\tilde{x}_2$, or to two nodes in $\tilde{x}_2$ and none in $\tilde{x}_1$, will also be updated, as if they had learned the intended conjunction. The effect is that whenever just one of $\tilde{x}_1$ or $\tilde{x}_2$ fires an allocation of new neuroids may be

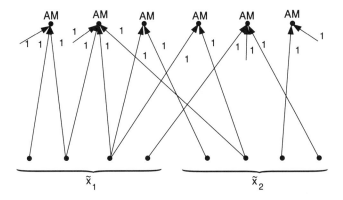

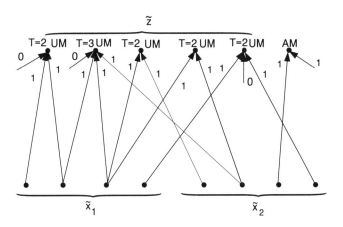

Figure 7.1. Illustration of the effects of Algorithm 7.1. The upper diagram shows a circuit fragment in its initial condition. The lower diagram shows it with all the weights, thresholds, and states updated. In each diagram the eight nodes at the bottom illustrate four $\tilde{x}_1$ nodes and four $\tilde{x}_2$ nodes. The nodes allocated to $\tilde{z}$ are the leftmost five among the six nodes in the top row. Note that the leftmost amongst these has no connection from any $\tilde{x}_2$ node.

made, and these will behave as replicas of $\tilde{x}_1$ or $\tilde{x}_2$. This problem of potentially uncontrolled sequences of redundant allocations of storage is a serious one. Fortunately, any one of several approaches may be taken to overcome it. For example, one could attempt to ensure that no two elements of any set representing one item have a common neighbor. This could be achieved either by making an appropriate initial assumption about the network or by executing an algorithm that somehow "burns out" these connections (i.e. makes them have weight zero) when $\tilde{x}_1$ and $\tilde{x}_2$ are first allocated.

We shall take a different approach here. The problem that needs to be overcome is that Algorithm 7.1 treats all pairs of nodes in $\tilde{x}_1 \cup \tilde{x}_2$ as having identical roles, which they do not. Those spanning the two sets should have a different effect from those entirely within only one of them. In other words, we need to break symmetry. The basic technique we will use for this here, and also subsequently, is *timing*. We will allocate to a peripheral the task of firing $\tilde{x}_1$ alone at one time, and firing $\tilde{x}_2$ alone at a later time. This peripheral can be thought of as an attentional mechanism that recognizes x_1 and x_2 as distinct, and is able to effect the firings of the corresponding nodes at distinct times.

Algorithm 7.2

Step 0: Prompt: $\tilde{x}_1$.

$$\{q_i = \text{AM}, w_i \geq 1\} \Rightarrow \{q_i := \text{AM1}, T_i := w_i,$$

$$\text{if} \quad f_j = 1 \quad \text{then} \quad w_{ji} := 2\}.$$

Step 1: Prompt: $\tilde{x}_2$.

$$\{q_i = \text{AM1}, w_i \geq 1\} \Rightarrow \{q_i := \text{UM}, T_i := T_i + w_i,$$

$$\text{if} \quad f_j = 0 \quad \text{and} \quad w_{ji} = 1 \quad \text{then} \quad w_{ji} := 0,$$

$$\text{if} \quad f_j = 0 \quad \text{and} \quad w_{ji} = 2 \quad \text{then} \quad w_{ji} := 1.\}$$

$$\{q_i = \text{AM1}, w_i < 1\} \Rightarrow$$

$$\{q_i := \text{AM}, T_i := \infty, w_{ji} := 1 \text{ for all } j.\}.$$

The result of the execution of Algorithm 7.2 is shown below in Figure 7.2.

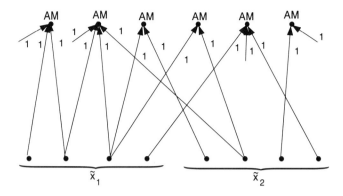

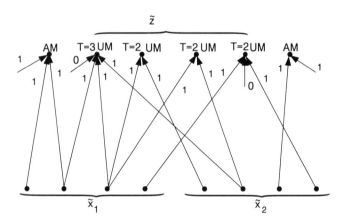

Figure 7.2. Illustration of the effects of Algorithm 7.2. The upper diagram shows a circuit fragment in its initial condition while the lower diagram shows it with all the weights, thresholds and states updated. In each diagram the eight nodes at the bottom illustrate four $\tilde{x}_1$ nodes and four $\tilde{x}_2$ nodes. The nodes allocated to $\tilde{z}$ are the central four nodes among the six illustrated in the top row.

In this algorithm the nodes $\tilde{x}_1$ fire first. All AM nodes adjacent to a member of $\tilde{x}_1$ go into state AM1, record the value of w_i in T_i, and adjust the weights on the edges from $\tilde{x}_1$ to 2, leaving the values of the others as 1. In the second step the $\tilde{x}_2$ nodes fire (and the $\tilde{x}_1$ nodes have ceased firing by default). Those AM1 nodes that are adjacent also to some $\tilde{x}_2$ node now go into unsupervised memory state (UM), update their threshold to equal the total number of $\tilde{x}_1$ and $\tilde{x}_2$ nodes to which they are adjacent, and update their weights so as to be one on edges coming from the $\tilde{x}_1$ and $\tilde{x}_2$ nodes, and to be zero from all the others. The remaining AM1 nodes, those not adjacent to $\tilde{x}_2$ nodes, revert to have the original mode and weights of available memory neuroids. We conclude, therefore, that this algorithm does indeed perform unsupervised memorization of 2-conjunctions as desired. We note also that only the $\tilde{x}_1$ and $\tilde{x}_2$ nodes fire in the course of the algorithm. The AM1 nodes do not undergo threshold firings since, when alternative transitions are available, as they are here, these are invoked instead. This allows T_i to be used to memorize a number rather than to represent a real threshold when a node is in state AM1. The avoidance of inessential firings can be important for an algorithm, since this will also minimize the side-effects. In this algorithm, for example, if the AM1 nodes had been allowed to fire then these would have caused a cascade of unintended storage allocations of a similar nature to the intended one.

Returning now to the problem of learning conjunctions of arbitrary length we observe simply that this can be done by means of the algorithm for 2-conjunctions, as long as the peripherals do appropriately more work. To learn $x_1 \wedge x_2 \wedge x_3$, for example, the peripherals would find some ordering on these three items, such as x_1, x_2, x_3. They would then supply x_1 and x_2 in a call of Algorithm 7.2 so that a new item $z_1 = x_1 \wedge x_2$ is memorized. Finally, they would supply z_1 and x_3 in a second call of that algorithm, so that $z_2 = z_1 \wedge x_3 = x_1 \wedge x_2 \wedge x_3$ is memorized. By supplying z_1 we mean that $\tilde{z}_1$ will be made to fire, which will be achieved by causing $\tilde{x}_1$ and $\tilde{x}_2$ to be fired simultaneously. It is easy to see that this strategy can be iterated so that conjunctions of any length n, not just 2 or 3, can be learned. The time needed for learning by this method would, of course, increase with n. Note that the order in which x_1, x_2, x_3 are presented does affect the structure of the circuit, which may be of the form $(x_1 \wedge x_2) \wedge x_3$ or $x_1 \wedge (x_2 \wedge x_3)$ or $(x_1 \wedge x_3) \wedge x_2$. When used for recognition, these circuits undergo threshold firing only, and x_1, x_2, x_3 can be presented simultaneously. Hence the order in which the three items are associated in the learning process will not, at this level, affect the predicate recognized.

In later chapters we shall discuss further the importance of general "systems" questions that arise whenever many instances of many functionalities coexist in the same neuroidal system. It will suffice here to observe that the specification of even a single algorithm for a single task poses some nontrivial problems if we allow that the algorithm be invoked more than once. Suppose that in this present case of unsupervised memorization, we use the algorithm to learn a sequence of concepts hierarchically. Suppose, in particular, that we are learning a conjunction, say of y_5 and y_6, where both constituents had been learned previously in terms of lower level items, such as $y_5 = y_1 \wedge y_2$ and $y_6 = y_3 \wedge y_4$. Whenever the peripherals need to fire $\tilde{y}_5$ and $\tilde{y}_6$ they will also have to fire the lower level nodes $\tilde{y}_1$, $\tilde{y}_2$, $\tilde{y}_3$ and $\tilde{y}_4$. There is, therefore, the danger that in addition to the intended conjunction, other unintended conjunctions, involving lower level items, such as $y_1 \wedge y_3$ will also be memorized as separate items. We shall return to general systems problems in Chapter 13. It will suffice here to note that, in this instance of pure conjunctions, there is nothing semantically wrong with memorizing unintended low level conjunctions in addition to the higher level ones. All the unintended conjunctions, such as $y_1 \wedge y_3 \wedge y_5$, are conjunctions of recognized attributes present in the input and it is not incorrect to memorize them in unsupervised node. We will take steps to control these unintended memory allocations, but only in order to minimize memory usage.

When several functionalities that relate to each other are supported within one system, then more particular requirements will need to be satisfied to ensure consistency. The functionality of memorization that we have been discussing is clearly related to the functionalities that correspond to the various possible methods of memory access. We shall return to this question of how information is retrieved from memory in Chapters 12 and 13.

We note, in conclusion, that as we introduce each new algorithm, we shall, for the sake of simplicity, examine it essentially in isolation in the first instance. Details not revealed in their descriptions will be assumed to be irrelevant or missing. In Figures 7.1 and 7.2, for example, it is assumed that the $\tilde{x}_1$ and $\tilde{x}_2$ nodes only fire when prompted, since no information about their state is given. It is also assumed that there are no edges among the $\tilde{x}$ nodes, among the $\tilde{z}$ nodes, or from the $\tilde{z}$ nodes to the $\tilde{x}$ nodes. These last assumptions (or close approximations to them) are guaranteed by the random graph assumptions. For example, if $r = 50$, $N > 10^8$ and the edge probability $p = (Nr)^{-1/2}$, then the probability that the neuroids representing two arbitrary items have even one of the r^2 potential edges between them present is remote.

Chapter 8
Supervised Memorization

8.1 Introduction

We formalize the problem of supervised memorization in a similar way to that of the unsupervised case. The main difference is that now the neuroids representing the item being learned are already allocated at the start of this process. This makes the problem both easier and harder than the unsupervised case. It is easier in the sense that no memory allocation mechanism is needed. On the other hand, it is harder in the sense that the constructed circuit now has to link nodes that have been previously selected and allocated, and the nodes for storing the target item can no longer be chosen for their convenient connectedness to other nodes.

More precisely, learning a conjunction $z = x_1 \land x_2 \land \cdots \land x_n$ now means that for previously fixed node sets $\tilde{x}_1, \tilde{x}_2, \cdots, \tilde{x}_n$, and $\tilde{z}$, a circuit has to be established such that whenever $\tilde{x}_1, \cdots, \tilde{x}_n$ all fire in the future, so will $\tilde{z}$. If, before this interaction, all incoming edges to $\tilde{z}$ had weight zero, then we can make the stronger claim that the only future condition that will make $\tilde{z}$ fire is the firing of $\tilde{x}_1, \cdots, \tilde{x}_n$ together. The more realistic case is one where the $\tilde{z}$ were allocated in the first place by unsupervised memorization and, therefore, some of their incoming edges already have positive weights. Then the previously stated conditions for firing $\tilde{z}$ are still sufficient but no longer necessary since mixtures of subsets of the nodes that partook in the unsupervised learning and of the new nodes $\tilde{x}_1, \cdots \tilde{x}_n$ may also suffice. Spurious behavior of this kind can be avoided by more complex mechanisms that ensure that either one of the intended inputs is sufficient to cause $\tilde{z}$ to fire, but partial mixtures of the two are not. However, as mentioned in §4.5, it is not clear whether the brain has mechanisms to prevent such mixed mode behavior.

We shall describe two algorithms for the basic task of supervised memorization. The first is simple and has the further advantage that it does not require edges to be bidirectional. The much more complex second algorithm assumes bidirectionality. Both algorithms, as described, implement a *graded* approximation of the true Boolean conjunction $x_1 \wedge \cdots \wedge x_n$. While most of the $\tilde{z}$ nodes will respond to exactly this conjunction, a small fraction, depending on the parameter μ ,described in §6.4, will be connected to representatives of only certain subsets of these n items and will respond to the firing of these subsets. Note that the graded quality here is provided by the random connectivity of the network. This quality may be desirable in cognitive modeling, to reflect the graded nature of concepts that has been detected in psychological experiments. At the expense of complicating the algorithms, the grading can be removed by having the nodes that fail to have the required connections detect and somehow eliminate themselves.

The main difference between what the two algorithms achieve lies in the fact that the first one allows some unintended interference effects from irrelevant inputs, though with provably small probability, while the second does not. In the second, bidirectional edges are used to reserve certain *relay* nodes for the exclusive use of just the one conjunction in hand. In contrast, relay nodes in the first algorithm are essentially shared among all the nodes that have edges directed toward them and realize the claimed functionality only with high probability. Both algorithms can be implemented on either random graphs or on random multipartite graphs. The diagrams suggest the latter case since that is the slightly easier one.

8.2 A Simple Algorithm

Our first algorithm for supervised memorization of a conjunction $z = x_1 \wedge \cdots \wedge x_n$ is illustrated in Figure 8.1. The $\tilde{z}$ nodes are initially in unsupervised memory state, UM, suggesting that they were initially allocated by some process such as Algorithm 7.2. The weights on the relevant edges incoming to the $\tilde{z}$ nodes can be assumed then to be zero, since that is how Algorithm 7.2 leaves all such edges, except for those coming from the nodes representing the features of the unsupervised memorization process. The initial values of the thresholds of the $\tilde{z}$ nodes are immaterial to the algorithm. The states of the $\tilde{x}$ nodes are also immaterial provided that they only fire when prompted, which is what we assume. Also, since the main advantage of the algorithm is that it does not require bidirectionality, we shall assume that there are no edges from $\tilde{z}$ back to the R nodes, or from the R nodes back to the $\tilde{x}$ nodes. In other words, we have a multipartite graph, such as shown in Figure 6.1(b), but in which all edges are directed upward.

Communication will take place through nodes in relay state R that have threshold one, and weight one on all incoming edges. A relay node is never updated, except that it undergoes threshold firing if at least one of its inputs is firing.

Algorithm 8.1 is then simply the following:

Step 0: Prompt: $\tilde{x}_1, \tilde{x}_2, \cdots, \tilde{x}_n, \tilde{z}$.

$$\{q_i = \text{UMF}\} \Rightarrow \{q_i := \text{UM1, if } f_k = 1 \text{ then } w_{ki} := 1\}.$$

Step 1: Prompt: $\tilde{x}_1, \tilde{x}_2, \cdots, \tilde{x}_n$.

$$\{q_i = \text{UM1}\} \Rightarrow \{q_i := \text{SM}, T_i := w_i\}.$$

This algorithm works as follows: When the nodes $\tilde{x}_1, \cdots, \tilde{x}_n$ fire, all the relay nodes to which they have directed edges undergo threshold firing essentially instantaneously. If the $\tilde{z}$ nodes in state UM are made to fire at the same time, then the edges to these $\tilde{z}$ nodes from the firing relay nodes will have their weights changed to one. Also each $\tilde{z}$ node will go into state UM1. At the next

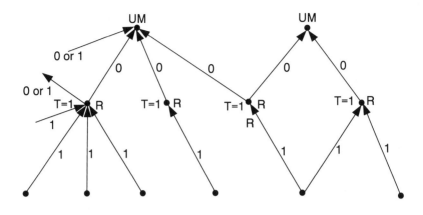

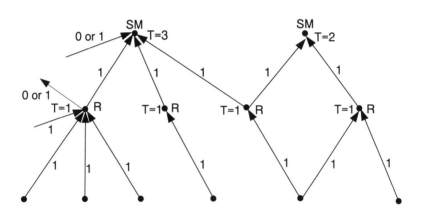

Figure 8.1. Illustration of the effects of Algorithm 8.1. The top diagram shows a circuit fragment in its initial condition. The lower diagram shows the states, weights and thresholds after the algorithm has been executed. In each case there are six $\tilde{x}$ nodes illustrated at the bottom of the circuit and two $\tilde{z}$ nodes at the top.

step the nodes $\tilde{x}_1, \cdots, \tilde{x}_n$ will be prompted again, thereby causing the same relay nodes to fire as before. The $\tilde{z}$ nodes will update their state to SM and their threshold to w_i, the number of edges incoming from firing relay nodes.

Many complications may arise when several algorithms are to be run compatibly on one network. To mention just one, we note that when a $\tilde{z}$ node first memorizes something in unsupervised mode and then goes on to memorize something in supervised mode, it is important that the latter process not destroy the effects of the former. In the present case changing the threshold may have such a deleterious effect. This problem can be overcome, however, by having the weights created in the second process adjusted so as to fit the old threshold. This can be done by replacing "$T_i := w_i$" in the transitions specified in Algorithm 8.1 by "if $f_k = 1$ then $w_{ki} := w_{ki}T_i/w_i$", where T_i denotes the threshold before the current update (i.e. the threshold acquired in the process of unsupervised memorization).

It remains to explain why the sharing of a relay node, among all the nodes from which there are connections to it, gives rise to unwanted effects only with negligible probability. For this purpose consider the r fixed nodes $\tilde{z}$, the nr fixed nodes $\tilde{x}_1, \cdots, \tilde{x}_n$, and also consider some kr other fixed nodes that form k disjoint sets $\tilde{y}_1, \cdots, \tilde{y}_k$. We want to show that the weights that have been changed to 1 by Algorithm 8.1 will not have the effect of causing many of the $\tilde{z}$ nodes to fire in the future when some spurious set of items $\tilde{y}_1, \cdots, \tilde{y}_k$ fire. In order to estimate this number, let j be any relay node, let p be the probability of an edge being present from any fixed node to any other, and let us assume that this probability is independent for the various pairs of nodes. Then the probability that this j is involved in such a damaging weight change is the probability that j is connected simultaneously to at least one member of $\tilde{z}$, at least one member of $\tilde{x}_1, \cdots, \tilde{x}_n$, and to at least one member of $\tilde{y}_1, \cdots, \tilde{y}_k$. This probability is

$$[1 - (1 - p)^r][1 - (1 - p)^{nr}][1 - (1 - p)^{kr}]$$

since $(1-p)^r$, $(1-p)^{nr}$ and $(1-p)^{kr}$ are the respective probabilities of each of these events not happening. Applying the Binomial Theorem to each component gives that the overall probability is $nkr^3p^3 + \mathrm{O}(p^4)$. We argue that if this quantity is less than about $1/N$, then this source of interference is negligible in the sense that the expected number of offending nodes j is less than one. Consider $p = (\mu/(rN))^{\frac{1}{2}}$ with $\mu = 4$, as suggested in §6.5, and $N = 10^{10}$. Then we need $Nnkr^3p^3 < 1$, if the $\mathrm{O}(p^4)$ term can be ignored, or $8nkr^{3/2} \leq 10^5$. If $r = 50$ then restricting $n, k \leq 5$ is sufficient.

We have shown that the expected number of relay nodes, intrinsic to the circuit for memorizing $z = x_1 \wedge x_2 \wedge \cdots \wedge x_n$, that a particular spurious conjunction $y_1, \cdots, y_k$ causes to fire, is at most one (i.e. $N \times 1/N$). This is small compared with the number of corresponding nodes in the intended circuit, which is about rn. A simple calculation shows that if this expected number is one, then it is extremely unlikely that a significant fraction of $\tilde{z}$ will be caused to fire if, say, $\tilde{z}$ is of size $r = 50$. The restrictions that were sufficient to guarantee all this were that n, the length of the conjunction, is at most five, and that k, the number of items firing spuriously at that instant is similarly bounded. The necessity of a bound of this latter kind is discussed further in §12.2 (and enunciated there as Principle 3). We note that at each instant a different set of k items may be firing. It is sufficient to show, as we have, that whichever set this is, the probability of unwanted side effects will be small at that instant. The phrasing of the proof was that the k items are first fixed, and the graph is randomly generated afterward. The meaning, however, is that the random graph is generated first, and the k items are chosen randomly afterward, but independently of the graph.

8.3   A Second Algorithm

The algorithm described below, Algorithm 8.2, is much more complex than the previous one, but guarantees that the interference effect analyzed above never occurs. It illustrates how on bidirected networks it is possible to control the behavior of subnetworks very precisely. The positive weight paths to UM nodes only carry influences that are intended. The issues that arise in this section are somewhat specialized and illustrate the programming possibilities for bidirected networks. We shall not have reason to revisit them.

As before, this algorithm has the target nodes $\tilde{z}$ initially in unsupervised memory state (UM). The $\tilde{z}$ nodes will establish communication with the $\tilde{x}$ nodes via some nodes initially in available relay (AR) state. Initially all weights on edges toward AR nodes have weight one, and those away from AR nodes have weight zero. The fact that the weights of the edges from the AR nodes to the UM nodes are initially zero is consistent with the scenario of the UM node having been previously trained by Algorithm 7.2, which leaves all such weights zero. The initial thresholds of the UM and AR nodes are immaterial to the algorithm and may be regarded as infinite. The effect of the algorithm is illustrated in Figure 8.2.

Algorithm 8.2

Step 0: Prompt: $\tilde{x}_1, \tilde{x}_2, \cdots, \tilde{x}_n, \tilde{z}$.

$$\{q_k = \text{ AR}, \ w_k \geq 2\} \Rightarrow \{q_k := \text{ AR1F }\}$$

$$\{q_i = \text{ UMF }\} \Rightarrow \{q_i := \text{ UM1}\}.$$

Comment: AR nodes connected to at least two of the prompted nodes will go to state AR1F at time 1. (This transition is not to be regarded as an instantaneous threshold transition since it is explicitly written out. Hence it takes unit time.) Also, at time 1 the $\tilde{z}$ nodes will be in state UM1F because they transited to state UM1 and are prompted at that time, but the $\tilde{x}$ nodes will have ceased firing, being assumed to be in a state that has that default.

Step 1: Prompt: $\tilde{z}$.

$$\{q_k = \text{ AR1F}, \ w_k \geq 1\} \Rightarrow \{q_k := \text{ AR2}\}$$

$$\{q_i = \text{ UM1F }\} \Rightarrow$$
$$\qquad \{q_i := \text{ UM2, if } f_k = 1 \text{ then } w_{ki} := 1\}$$

$$\{q_k = \text{ AR1F}, \ w_k < 1\} \Rightarrow \{q_k := \text{ AR }\}.$$

Comment: Those AR1F nodes that are connected to $\tilde{z}$ will go to state AR2, the edges from the former to the latter being given weight 1. The remaining AR1F nodes, those not connected to any $\tilde{z}$ node, will revert to the pristine AR state, while the $\tilde{z}$ nodes

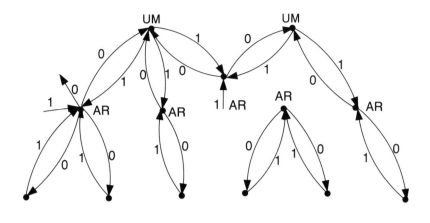

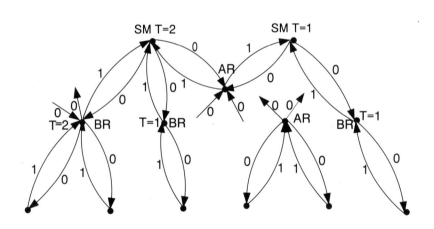

Figure 8.2. Illustration of the effect of Algorithm 8.2. The top diagram shows a circuit fragment in its initial condition. The lower diagram shows the changes to the weights, states, and thresholds after the algorithm has been executed. In each diagram there are six $\tilde{x}$ nodes illustrated at the bottom of the circuit and two $\tilde{z}$ nodes at the top.

will proceed to state UM2. (Note that by having state AR1F, we are using the firing status of a node to communicate information between steps. We could avoid this by having the $w_{ki} := 1$ updates done in an extra step having $\tilde{x}_1, \cdots, \tilde{x}_n$ as the prompt, and added between the current Step 2 and Step 3.)

Step 2: Prompt: $\tilde{x}_1, \cdots, \tilde{x}_n$.

$$\{q_k = \text{AR2}, w_k \geq 1\} \Rightarrow \{q_k := \text{BR}, T_k := w_k,$$
$$\text{if } f_l = 0 \text{ then } w_{lk} := 0\}$$

$$\{q_i = \text{UM2}\} \Rightarrow \{q_i := \text{UM3}\}$$

$$\{q_k = \text{AR2}, w_k < 1\} \Rightarrow$$
$$\{q_k := \text{AR}, w_{lk} := 0 \text{ for all } l\}.$$

Comment: The AR2 nodes connected to some $\tilde{x}$ nodes have T set to the number of such nodes. All weights to them from other nodes are set to zero, and they themselves go to busy relay (BR) state. The AR2 nodes not connected to any $\tilde{x}$ nodes become isolated by having all incoming weights set to zero. The $\tilde{z}$ nodes progress to state UM3.

Step 3: Prompt: $\tilde{x}_1, \cdots, \tilde{x}_n$.

$$\{q_i = \text{UM3}\} \Rightarrow \{q_i := \text{SM}, T_i := w_i\}.$$

Comment: By threshold firings all the BR nodes adjacent to the prompted $\tilde{x}$ nodes are caused to fire. The $\tilde{z}$ nodes progress to supervised memory state (SM). Their thresholds are adjusted to the number of BR nodes through which they are connected to some $\tilde{x}$ nodes.

As in the unsupervised case the issue arises as to how these algorithms fare when learning is hierarchical, or, in other words, when to make $\tilde{x}_1, \cdots, \tilde{x}_n$ fire it is also necessary to make some lower level items fire. The observation made in the previous case, that no semantic problems occur if all the items are pure conjunctions, holds here also. Every item that is true of an input, whether at a higher or lower level in the hierarchy, is a valid attribute of the input, and it is valid, though perhaps not efficient, to form the conjunction of all of them when memorizing the input as a conjunction.

Chapter 9
Supervised Inductive Learning

9.1   Introduction

As explained earlier, we find it useful to divide knowledge acquisition mechanisms into two categories. In the first the acquisition mechanism is such that little room is left to question the validity of the acquisition process. Examples of acquisition methods of this first kind are: inheriting an algorithm at birth, memorizing the text of a program or recipe given to one explicitly, memorizing the features of an individual or of an event, and arriving at knowledge by logical deduction from knowledge already available. Whatever problems of computational implementation these processes may pose, none of them raises philosophical difficulties concerning the rational defensibility of the outcome.

This chapter is devoted to knowledge acquisition mechanisms of the second kind where the process itself is no longer immediately defensible. While this definition may encompass a broad range of possibilities, there is an instance of it that we consider to be paradigmatic and this is *learning by example*. Here the teacher has a predicate or *concept* in mind, such as that of a chair. The learner is presented with a number of objects each identified by the teacher as either being an example of the concept or as not being an example of it. The task of the learner, in its most operational form, is to acquire the capability of categorizing previously unseen objects as being or not being examples of the same concept (e.g. chairs or not chairs). The process of knowledge acquisition by such indefensible means is usually called inductive learning or often, in our case, just learning, for short.

Our definition of inductive learning can be viewed as a negative one: any mechanism that is not defensible in a straightforward way, but has some validity, suffices. In general, inductive methods share the property that the learner acquires capabilities that are not logically implied by the information presented. Clearly these mechanisms embrace a rich variety of phenomena. The task to be learned may be that of learning new concepts, or that of increasing some measure of performance at some task that has been mastered qualitatively. Also, the interaction with the world may take many forms besides the paradigm of learning from examples.

Induction has a paradoxical aspect in its very definition. How is it possible to abstract more useful information from information presented than is logically implied in it? Philosophers have given much thought to this question, and have often emphasized both its centrality and its problematic character. Hume described induction as a process in which the regularities observed through experience lead one to acquire habits of what to expect in the future. In an extreme case if occurrences of a situation in the past always had identical outcomes, then one expects a reoccurrence to lead to the same outcome the next time. His statement clearly has much verisimilitude, but it leaves unresolved the exact nature of the regularities that are useful for induction in general. Because of these difficulties some philosophers in the twentieth century have questioned the very existence of induction in the classical sense (N. Goodman 1983).

In our view these philosophical questions can be finessed by regarding induction more as a natural phenomenon and less as a logical paradox. As in other areas of science it is then sufficient to model and explain some significant aspects of this phenomenon. We do not need to arrive at a formal definition that accounts for all senses in which the word induction is commonly used.

The aspects of induction that are most promising for scientific investigation are those that are most mundane, mechanical, and experimentally repeatable. Every day millions of children learn new words for artifactual concepts, such as pieces of furniture, and it seems implausible that these concepts are somehow already preprogrammed in the genes. Furthermore, simple inductive phenomena appear to go back a long way in evolutionary development. As mentioned in Chapter 4, there are many remarkable experiments that show that pigeons and other animals have substantial powers of generalization. We may view induction, therefore, as a concrete reproducible phenomenon in the same spirit as a physical phenomenon might be viewed in mechanics. We need to be aware, however, that there is a wide range of other phenomena, such as scientific discovery or artistic creativity, that may have features in common with this view of induction, but that also seem to have additional aspects. These phenomena have proved much more difficult to specify or to reproduce under experimental conditions, and we make no attempt at considering them here.

9.2 Pac Learning

Since learning is a central aspect of cognition, we need to decide on what view to take of it. There is a field of computer science, called *computational learning theory*, that is concerned with modeling and understanding the learning capabilities of computational systems. A first aim of this field is to provide plausible specifications of what a system that learns inductively can be expected to achieve. For a mathematically well understood task, such as integer multiplication, it is easy enough to specify what a program for accomplishing it should do. We need to have similarly clear criteria for learning if we are to evaluate whether various proposed mechanisms for realizing it are effective. For example, if we were to purchase a home robot that is advertised as adaptable or having learning capabilities, what kind of guarantee of performance should we look for?

It seems that any acceptable criterion of inductive learning has to reconcile two opposing requirements. It has to be strict enough that, if satisfied, induction in a useful and reliable sense is indeed achieved. On the other hand it should not be so strict that it is impossible to deliver that level of performance by any feasible computational mechanism.

The criterion of learning that we shall follow here is what is called *distribution-free* or *probably approximately correct* learning (*pac* for short). An informal rendering of this on the guarantee of the stipulated home robot would read something as follows: whatever home you take this robot to, after sufficient training on some tasks it will behave as expected most of the time, as long as the general conditions there are stable enough. To make this informal statement into a usable criterion, some quantitative constraints are needed in addition. First, the number of training sessions required should be reasonable, as should the amount of computation required of the robot to process each input at each such session. Second, the probability that the robot fails to learn because the training instances were atypical should be small. Lastly, the probability that, even when the training instances were typical, an error is made on a new input should be small. Furthermore, in the last two cases the probability of error should be controllable in the sense that any level of confidence and reliability should be achievable by increasing the number of training instances appropriately.

This intuition can be captured formally for several learning contexts. In the simplest case, of learning from examples, this can be done as follows: All the examples, whether during training or testing, are considered to be drawn randomly and independently from one probabilistic distribution, one that describes the probability of the various examples occurring in the real world. It is not necessary for the learner to understand the complexities of the distribution, but only to classify reliably new examples drawn from that distribution. Understanding the distribution would correspond to having complete information about the world. Classifying new examples corresponds to merely coping with it. The idea is that simple strategies may suffice to cope with a world that is too complex for us to fully understand. More importantly, these simple strategies may be learnable effectively through experience with the world.

Details of how such a theory can be made precise can be found elsewhere in the literature[22]. Briefly, all the quantitative constraints can be stated, in the first instance, as requiring that the associated quantities grow only polynomially (i.e. as a fixed power of n such as n^2 or n^3) as opposed to faster, such as exponentially (i.e. 2^n). The parameter n represents the relevant measures of the problem. These include the size of the program or strategy being learned, and the inverse of the probability of error that is to be tolerated. A class of programs is *learnable* from examples in this sense if the number of training inputs and computational steps needed is bounded as a polynomially growing function of the relevant parameters. This framework can be used in richer settings also, such as when the learner can ask the teacher questions. We shall restrict ourselves here to the case of learning by example. We shall not need to go further into the technicalities of learning theory. It will be sufficient that the particular algorithms we implement on the neuroidal model be such that they are known to be learnable in this sense.

It turns out that the classes of representations that are learnable, even when the learning process is allowed to run on a general purpose computer as is allowed by the pac learning model, is severely limited. Examples of simple classes that can be learned include Boolean conjunctions (e.g. $x_1 \wedge x_5 \wedge \bar{x}_7$) and Boolean disjunctions (e.g. $x_1 \vee \bar{x}_3 \vee x_8$). Because of their simplicity these classes are more appropriately called classes of *knowledge representations* rather than of programs, since the computational aspect of evaluating these Boolean forms is trivial compared with the complexities that may be found in arbitrary programs.

An important class that is not currently known to be learnable is *disjunctive normal form* (or DNF for short). As described in §4.2, this consists of expressions that are of the form of a disjunction over terms that are conjunctions themselves. Thus $x_1 \bar{x}_2 x_3 \vee x_1 x_2 \vee x_2 x_4 x_7$ is such a DNF expression where conjunctions such as $x_1 \wedge \bar{x}_2 \wedge x_3$ are abbreviated in standard product notation as $x_1 \bar{x}_2 x_3$. This appears to be a most natural generalization of simple conjunctions from the viewpoint of modeling human concepts. It can express the idea that examples of a concept fall into a number of somewhat distinct categories, each corresponding to one of the conjunctions.

When discussing inductive learning we have a *hierarchical* context in mind. If we wish to learn DNF formulae, but do not have an algorithm for learning these directly, we can nevertheless attempt to learn these in stages. For example, to learn $x_1 x_2 \lor x_2 x_3$ we could first learn the simple conjunctions $x_1 x_2$ and $x_2 x_3$ separately in some fashion. Having learned these we can learn the DNF formula using an algorithm for simple disjunctions. Of course, when learning hierarchically in this way more is required of the teacher or environment than in the simplest case of learning by example. Somehow the subconcept $x_1 x_2$ must be learned separately in supervised or unsupervised mode. In the former case, for example, a teacher may have to teach the name of this subconcept in unsupervised memorization mode and then identify positive or negative examples of it so that it is learned in supervised mode inductively. Alternatively, this subconcept may be learned in unsupervised mode either by memorization or by correlational learning.

In this context, learning theory can be thought of as defining the granularity with which learning can proceed without intervention from the outside. The largest classes of programs that are learnable represent the largest chunks of information that can be learned feasibly without their having to be broken up into smaller chunks.

Our current state of knowledge suggests that this granularity is small and that there are severe restrictions on the classes of knowledge representations that are learnable in polynomially many computational steps on general purpose computers. If we restrict ourselves to learning on our neuroidal model in at most ten or a hundred steps per input, then we are tying our hands behind our backs. Somewhat surprisingly, it appears that some of the most natural classes currently known to be learnable on the general model are not significantly restricted when we constrain ourselves in this particular way. In the next three sections we shall describe known pac learning algorithms for three classes: conjunctions, disjunctions, and linear threshold functions. We shall show that they are implementable on the neuroidal model.

One plausible approach for dealing with DNF representations on a general purpose computer is the following: First, restrict the constituent conjunctions to those that are somehow learnable in unsupervised mode, and then learn the DNF as a supervised disjunction over these.[23] The unsupervised learning may consist of identifying sets of variables that are found to be true simultaneously in natural examples with unexpectedly high probability and memorizing their conjunctions. The supervised learning consists then of finding a subset of these conjunctions whose disjunction approximates the function being learned. It turns out that this strategy remains essentially feasible in the more restricted neuroidal model! This gives us some confidence that the bounds of learnability, at least for the style of knowledge representation that we consider here to be relevant to cognition, have some robustness to changes of computational model. This is exactly what is required for a robust theory.

9.3   Learning Conjunctions

Suppose that a system is learning a recognition rule that is known *a priori* to be a conjunction of a subset of the Boolean predicates $x_1, \cdots, x_n$, for each of which a recognition algorithm is already available to the system. Suppose that the true rule is $x_2 x_7 x_9$. This means that every possible positive example will have these three attributes true and, conversely, that whenever all three are true then the example is a positive one. To induce such a rule from examples there is an obvious and classical method called *elimination*. In this we consider a set K that initially includes all n of the x_i predicates. Each time we see a positive example where some x_j is false, we eliminate this x_j from K, since any such single example is sufficient evidence to confirm that x_j cannot occur in the correct rule. At each stage in this process the hypothesis is made that the correct rule is the conjunction of all the predicates that have not been eliminated from K.

This simple algorithm has some strong properties. First, it is clear that it uses only positive examples and no negative ones. Second, it will never misclassify a real negative example as positive. At each step when a predicate is eliminated from K the examples that are newly accepted as positive by the altered hypothesis, namely the conjunction of the members of K, are guaranteed to be truly positive since the truth of the eliminated predicate cannot have been a condition necessary for positivity. If the true rule is $x_2 x_7 x_9$, then at any stage of this process these three predicates will remain members of K. Hence the only possible source of error in this algorithm is that some positive examples are not recognized as such by the current conjunction because the latter contains some extra predicates that are not satisfied by every positive example. For example, K may contain x_4 in addition to x_1, x_7, and x_9. It is possible to show, however, that if the examples are drawn randomly from some arbitrary probability distribution — the arbitrariness reflecting the complexities of the real world — and if the hypothesis is to be tested on a random example from the *same* distribution, then this source of error is well controlled. The essential point is that if there are predicates, such as x_4, that the examples ought to eliminate but fail to then, unless the examples drawn were atypical of the distribution, it must be the case that most positive examples do have these variables true. Hence retaining them in the hypothesis will have the effect of misclassifying only a few rare positive examples for which these predicates are false.[24] This probably approximately correct, or pac, behavior is acceptable, and is the best that can be hoped for.

Our first algorithm for implementing the inductive learning of conjunctions will be this elimination algorithm. We have items $x_1, \cdots, x_n$ and a target item z. In the ith interaction the ith positive example will be presented. The peripherals will prompt simultaneously $\tilde{z}$, as well as every $\tilde{x}_j$ such that x_j is true for this i^{th} example. The task for the learning algorithm is to ensure that after i such examples, a connecting circuit will exist between the $\tilde{x}$'s and the $\tilde{z}$, that will behave as follows: On a future input $\tilde{z}$ will be caused to fire by threshold firings through this circuit if and only if all those $\tilde{x}_j$ that fired in all the i positive examples already seen are firing. This is another way of saying that the circuit implements the elimination algorithm. (As in the algorithms described in the previous chapter the "only if" part of the claim applies only to the effect on $\tilde{z}$ of the *new* circuit that is being set up. In general, the $\tilde{z}$ nodes may also fire under additional conditions imposed by other circuitry previously established. For example, they may respond to inputs corresponding to their name learned earlier in unsupervised mode.)

We assume that some nodes have the capacity to learn conjunctions. On having succeeded these will be in *supervised conjunctive* (SC) state. We assume that initially they are allocated by some unsupervised process and hence we call their state UC for *unsupervised conjunctive*. They have the potential to learn a conjunction inductively exactly as a UM node has the potential to memorize a conjunction. Initially all the incoming weights have value one, and the threshold is effectively infinite.

At the presentation of the first example, any algorithm for supervised conjunctions in the style of Algorithms 8.1 or 8.2 may be used to learn the conjunction of all the attributes true in this first example. The only difference needed is that the $\tilde{z}$ nodes, rather than going through the state sequence UM $\Rightarrow$ UM1 $\Rightarrow$ SM or UM $\Rightarrow$ UMF $\Rightarrow$ UM1F $\Rightarrow$ UM2 $\Rightarrow$ UM3 $\Rightarrow$ SM, as in these previous algorithms, go instead through a renamed but otherwise identical sequence UC $\Rightarrow$ UC1 $\Rightarrow$ SC or UC $\Rightarrow$ UCF $\Rightarrow$ UC1F $\Rightarrow$ UC2 $\Rightarrow$ UC3 $\Rightarrow$ SC. What the first stage achieves is effectively to assign to K the set of all variables that are true for the first positive example.

Each successive stage will be prompted by the presentation of a further positive example, and will result in the execution of Algorithm 9.1. Each such execution will have the effect of eliminating any member of K that is false for that current example. To achieve this, it will simply set to zero the weight of any edge coming to the SC node from relay nodes that are not made to fire by the example. Also the threshold of the SC node will be reduced at each stage to compensate for the reduced weight sum of its edges. Note that w_i will have unit contributions from every w_{ji} such that j represents one of the items remaining in K. It will have zero contribution from every other j.

Algorithm 9.1

Step 0: Prompt: $\tilde{z}, \{\tilde{x}_j | x_j$ true for current example$\}$.

$$\{q_i = \text{SCF}\} \Rightarrow$$
$$\{q_i := \text{SC}, \ T_i := w_i, \ \text{if } f_k = 0 \text{ then } w_{ki} := 0\}.$$

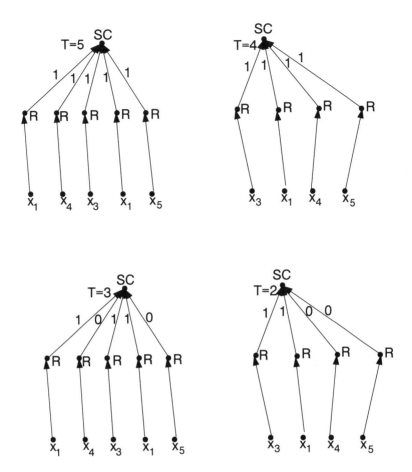

Figure 9.1. Illustration of the elimination algorithm for learning conjunctions, in the case that Algorithm 8.1 is used to process the first example. The top diagram shows a circuit fragment after a first example has been processed, in which $x_1 = x_3 = x_4 = x_5 = 1$ and all other $x_i = 0$. The lower diagram shows the result of subsequently executing Algorithm 9.1 for a second example in which $x_1 = x_3 = x_6 = 1$ and all other $x_i = 0$.

If a relay node that is connected to a $\tilde{z}$ node is connected to more than one $\tilde{x}$ group, say $\tilde{x}_j$ and $\tilde{x}_k$, then eliminating its connection to $\tilde{z}$ because $\tilde{x}_j$ fails to hold for a positive example would eliminate $\tilde{x}_k$ also, possibly mistakenly. Fortunately, this happens very rarely under the random graph assumptions made. By a calculation similar to the one used in §8.2 it can be deduced that the probability that any fixed relay node is connected to $\tilde{z}$ and to $\tilde{x}_j$ and $\tilde{x}_k$, for some j and k $(1 \leq j < k \leq n)$, is

$$[1 - (1-p)^r][1 - (1-p)^r][1 - (1-p)^r][n(n-1)/2]$$

$$\leq \frac{n^2}{2} r^3 p^3 + O(p^4)$$

$$\leq \frac{n^2}{2} \frac{r^{3/2} \mu^{3/2}}{N^{3/2}} \quad \text{if } p = (\mu/(rN))^{1/2}$$

$$\leq \frac{n^2}{25} \cdot \frac{1}{N} \quad \text{if } r < 100, \mu = 4 \text{ and } N = 10^{10}$$

$$\leq 1/N \quad \text{if } n^2 \leq 25.$$

We conclude that if $n \leq 5$ then the expected number of such relay nodes is at most one, and hence that the elimination algorithm works on most of the r nodes of $\tilde{z}$ as intended with high probability.

As an alternative, if we wish to avoid having such a graded response, then we can use Algorithm 8.2 and adapt it so that the offending relay nodes are identified and isolated. For example, when processing the first example, we could add an extra step before the current Step 3 of Algorithm 8.2:

Prompt: $\tilde{x}_1, \cdots, \tilde{x}_n$.

$$\{q_k = \text{ BR}, \ w_k \geq 2\} \Rightarrow \{w_{lk} := 0 \ \text{ for all } \ l\}.$$

This will ensure that relay nodes made busy by more than one $\tilde{x}$ node will be effectively eliminated, before the last, now fourth, step of the algorithm is executed.

As a general remark, we note that it is very desirable that a learning algorithm be *resilient* to errors in the data. In the case of supervised learning this means, for example, that if a small fraction of the examples are presented to the NTR as exemplifying the item being learned when they do not, and vice versa, then the learning algorithm should still succeed. The simple elimination algorithm just described is not robust, since a variable may be mistakenly eliminated from K once and for all, by just one example that is mistakenly labelled as being negative. The algorithm can be adapted, however, to be resistant to a small error rate by having a variable removed from K only if in a sufficient fraction of the positive examples this variable is false.[25]

A further general consideration in any inductive algorithm is that its behavior be appropriate for all interactions after the initial one. In the algorithm described above the precondition that the SC state be firing is intended to express the situation that this node is being prompted to indicate that the current example is positive. As specified the algorithm would be invoked also, however, by any input that causes the SC node to undergo threshold firing by virtue of all x nodes corresponding to the variables in K firing. Fortunately, the algorithm would then make no updates, and hence remains correct.

9.4   Learning Disjunctions

When learning disjunctions we assume that the examples and counterexamples presented are consistent with a Boolean disjunction over a subset of previously recognized Boolean predicates $x_1, \cdots, x_n$. If the rule is $x_1 \vee x_5 \vee x_{11}$, then all positive examples have at least one of x_1, x_5 or x_{11} true, while all negative examples have all these attributes false.

The elimination algorithm described for conjunctions has a precise dual, in which learning takes place from negative examples alone. We start by considering the disjunction of the set K of all the variables. Each negative example is then taken in turn and any x_i that is true in any negative example is eliminated from K since its truth is then known not to be sufficient to guarantee an example being positive. After any number of examples the rule that is hypothesized is the disjunction of the members of K that remain. This algorithm can be analyzed in a similar manner to its dual for conjunctions, and shown to increase in reliability in the pac sense in the same way as the number of examples seen increases.

Implementing this algorithm for disjunctions in a neuroidal network involves some new complications. As in the case of conjunctions it is impractical to start eliminating with a starting situation that has K containing all the predicates in the NTR. For conjunctions we had the easy solution of taking the initial members of K to be all the predicates that are true in the first example, since the predicates occurring in the true rule must be a subset of these. For disjunctions there is no comparably simple solution. If the true rule is $x_1 \lor x_5$, it may be that the first example has x_5 and some other predicates true, but x_1 false.

The algorithm we propose is the following. After each example we have a circuit as in Algorithms 8.1, 8.2 or 9.1, except that each target node has threshold one, so that it recognizes a disjunction rather than a conjunction. When a negative example is seen then some variable may be eliminated from the disjunction by setting some weights of the target node to zero. When a positive example is seen that is already covered by the current value of K then no updates are made. If the positive example does not make positive any member of K, then the attributes that are true in it are added to K. This augmentation of the circuit is carried out by running a version of Algorithm 8.1 or 8.2 on it as described below. Thus every attribute that is ever witnessed as occurring in even one positive example that is not already accounted for by the current K, is given consideration as a potential member of the disjunction being learned.

The sequence of interactions that result in learning a disjunction by elimination will first need, therefore, the input of a positive example. For dealing with this first example our algorithm will be essentially the same as Algorithm 8.1 or 8.2 except now the state sequence executed will be renamed UD $\Rightarrow$ UD1 $\Rightarrow$ SD or UD $\Rightarrow$ UDF $\Rightarrow$ UD1F $\Rightarrow$ UD2 $\Rightarrow$ UD3 $\Rightarrow$ SD, where D denotes disjunctions. The only additional modification is that when the final state SD is entered, the threshold is made one, rather than w_i, so that a disjunction is realized. For example, for Algorithm 8.1 the last step would become:

$$\{q_i = \text{UD1}\} \Rightarrow \{q_i := \text{SD}, \ T_i := 1\}.$$

How do we adapt the algorithm so that it can add predicates to K that are present only in positive examples subsequent to the first? It suffices to use an algorithm that is essentially the same as that used for the first example, except that it starts in state SD and goes through the sequence SD $\Rightarrow$ SD or SD $\Rightarrow$ SDF $\Rightarrow$ SD1F $\Rightarrow$ SD2 $\Rightarrow$ SD3 $\Rightarrow$ SD in analogy with Algorithm 8.1 or 8.2, respectively. We need to make two observations about this part of the algorithm. First, since no edge of weight 1 in an SD node will be made zero, no items from the current set K, over which a disjunction is being computed, will be removed from it. Second, any attribute in the new example that is not already present in the disjunction will be added as a disjunct. We note that this algorithm only needs to be invoked for those positive examples that fail to make $w_i > 0$ for the SD node.

It remains to say how negative examples are treated. In this case the supervised disjunctive nodes SD are prevented from firing by the prompt. We wish to make zero all edges to them from nodes that are firing, since these correspond to items that are clearly not sufficient to make the disjunctive concept being learned true. To achieve this it is sufficient to add the transition:

$$\{q_i = \text{SD}\} \Rightarrow \{\text{if } f_k = 1 \text{ then } w_{ki} := 0\}.$$

This completes our description of a simple elimination algorithm for learning disjunctions. Since any positive example may add spurious predicates to the disjunction, reliable learning can only be considered to have taken place if enough negative examples have been seen after the last positive one to ensure that all harmful predicates have been eliminated with high probability. Hence this algorithm may be thought of most simply as follows: First, enough positive examples are considered that all the predicates appearing in the true disjunction are seen at least once. Subsequently sufficiently many negative examples are seen that spurious items are eliminated from the disjunction with high probability. This two phase training sequence can be repeated, of course, any number of times.

One further point about the algorithm is that, as described, it will reintroduce an attribute that was previously eliminated by a negative example, if it occurs again in a positive one. Eliminating items previously eliminated has the same cost in further negative examples as newly added items not previously eliminated. Hence the degradation in performance is not of a radical nature. At the expense of making the algorithm more complex, we can modify it, however, so that eliminated attributes are never reintroduced.

9.5 Learning Linear Threshold Functions

A *linear threshold function* over real variables $x_1, \cdots, x_n$ is determined by a set of real parameters $u_1, \cdots, u_n$ and θ. Its value is defined to be one or zero according to whether the values $x_1, \cdots, x_n$ satisfy

$$u_1 x_1 + u_2 x_2 + \cdots + u_n x_n \geq \theta$$

where θ is called the threshold parameter. In cases where the coefficients u_i are allowed to be negative, an equivalent formulation of the criterion is

$$u_1 x_1 + u_2 x_2 + \cdots + u_{n+1} x_{n+1} \geq 0,$$

since we can choose x_{n+1} to be a predicate that always equals one, and u_{n+1} to equal $-\theta$. In the context of neuroids, we can associate with each neuroid a linear threshold function, interpret u_j as the weight of the edge coming to that neuroid from neuroid j, and interpret x_j as the firing status f_j of j.

When $x_1, \cdots, x_n$ are restricted to have values one or zero, then threshold functions can be viewed as Boolean functions and become candidates for representing items in a neuroidal net. Linear threshold functions are quite expressive as Boolean functions, being able to represent conjunctions, disjunctions, and more. A conjunction $x_1 \wedge x_3 \wedge x_4$ would correspond to $\theta = 3, u_1 = u_3 = u_4 = 1$ and the remaining $u_j = 0$. In other words,

$$x_1 + x_3 + x_4 \geq 3.$$

A disjunction $x_1 \vee x_3 \vee x_4$ would be expressed as

$$x_1 + x_3 + x_4 \geq 1.$$

Finally, a threshold function expressing the condition that at least k of a set of variables are true can be expressed as

$$x_1 + x_2 + \cdots + x_n \geq k.$$

In any of these a negation $\neg x_i$ can be represented by substituting $(1 - x_i)$ for x_i (e.g. in this last threshold function making $u_i = -1$ and θ one smaller than it would be otherwise). In particular, $x_1 \wedge (\neg x_3)$ would become $x_1 + (1 - x_3) \geq 2$, or

$$x_1 - x_3 \geq 1.$$

Note that one can avoid the arbitrary use of negative coefficients by introducing new variables. For example, in this case we could introduce a variable, say x_4, such that $x_4 \equiv \neg x_3$, and then use the inequality $x_1 + x_4 \geq 2$. In the case, for example, that x_3 is an item expressing the predicate "green" and has some neuroids representing it, there would be further neuroids representing the item "not green."

Linear threshold functions are of interest to us here for at least three reasons. First, it turns out that despite their expressive power being greater than that of conjunctions and disjunctions, they are still learnable though not quite as easily. Second, they capture the functional dependence of our neuroids on their inputs, and hence can be represented easily and naturally in our networks. Lastly, there are algorithms for learning them in significant special cases, where the update algorithms themselves are ideally suited to implementation on neuroids.

Interest in learning algorithms for linear threshold functions goes back several decades, in particular to the discovery (F. Rosenblatt 1962) of the very elegant *perceptron* algorithm. It can be applied in the following context. We assume that each example is presented as a vector $\underline{e} = (e_1, \cdots, e_n)$ where each e_j equals 0 or 1. We further assume that there exists a linear inequality

$$u_1 x_1 + u_2 x_2 + \cdots + u_n x_n \geq 0 \qquad (9.1)$$

that expresses the true predicate being learned. In other words, a vector $\underline{e} = (e_1, \cdots, e_n)$ is a positive or negative example of the predicate according to whether inequality (9.1) is or is not satisfied when e_j is substituted for x_j for every j. What the algorithm does is the following. It makes an initial hypothesis

$$v_1 x_1 + v_2 x_2 + \cdots v_n x_n \geq 0 \tag{9.2}$$

by making an arbitrary choice for each of the coefficients $v_1, \cdots v_n$. Choosing them all to be zero is an appropriate choice. It then considers each example $\underline{e}$ presented in turn. If the hypothesis gives the same classification for $\underline{e}$ as the true classification (i.e. if $\underline{e}$ either satisfies both of (9.1) and (9.2), or neither of them) then the algorithm leaves the hypothesis unchanged. Otherwise, if $\underline{e}$ is a positive example but does not satisfy the hypothesis (9.2) then e_j is added to v_j for every j. (Intuitively, this will tend to have the effect that the hypothesis is pushed toward being satisfied on future presentations of the same example $\underline{e}$). Conversely, if $\underline{e}$ is a negative example and does satisfy (9.2) then e_j will be subtracted from v_j for every j.

This algorithm is in a family of related algorithms that we characterize as *local*, and are well suited to implementation on neuroidal circuits. The basic property they share is that the update to each v_j is a function of quantities on which updates to neuroidal weights are allowed to depend. In particular we define an algorithm for learning $\sum v_j x_j \geq \theta$ to be local if, when an example with attribute values $x_1, \cdots, x_n$ is presented, the update to each v_j depends only on (i) the value of $\sum v_j x_j$, (ii) the value of x_j, (iii) the value of v_j, and (iv) whether the current example is positive or negative. If we implement the threshold function at neuroid i and the variables $\{x_j\}$ at appropriate neighboring nodes $\{j\}$ with $w_{ji} = v_j$, then condition (i) refers to w_i, condition (ii) refers to the firing status f_j of neighbor j, condition (iii) refers to the previous value of w_{ji} and condition (iv) refers to the correct classification of the example, which can be represented at i by the state q_i. The weight transition function λ, as we defined it for neuroids in Chapter 5, is allowed to depend on all four of these quantities. Note that local algorithms can be viewed as capturing the power of a single neuroid in the case that the mode has just two allowed values, (indicating whether the example seen is positive or negative).

We can implement local algorithms as follows. The target nodes will be initially in SL (supervised local) state. We need to set up a connecting circuit between the relevant nodes $\tilde{x}_1, \cdots, \tilde{x}_n$ and the target nodes in the manner of Algorithms 8.1 or 8.2. Hence the first phase will be to supply a number of examples (whether positive or negative) that between them contain all the variables necessary for a linear threshold function to exist for the desired predicate. Exactly as in the algorithm for supervised disjunctions, a connecting circuit from the items representing these variables to the target nodes will be formed first.

In the second phase the local algorithm is executed directly as each further example is presented. The weights $\{w_{ji}\}$ of the nodes originally in state SL correspond to the coefficients $\{u_j\}$ of the threshold function being learned. If a target node is connected to more than one member of an $\tilde{x}_j$ (i.e. via distinct relay nodes), then we have the slight complication that the set of relevant items $\{x_j\}$ effectively contains repetitions. This is no problem, however, for the perceptron or any other reasonable local algorithm, since if the x_j occurs with coefficient u_j in the true threshold function but now has k incarnations, say $x_j^{(1)}, \cdots, x_j^{(k)}$, then we still have ample choices for the coefficients of these incarnations, such as having the value u_j/k for each of them. For similar reasons, these algorithms are robust against irrelevant attributes since these can be given coefficient zero. Thus a relay node connecting a target node to nodes from more than one $\tilde{x}_j$ will do no harm either since, at worst, it will be treated as an irrelevant attribute by the learning algorithm. The generic reasons that make the learning graded, however, do remain. Each target node has connections to the relevant $\tilde{x}_j$ only with high probability. Hence a few of the target nodes may fail to learn the correct function.

Rosenblatt's perceptron algorithm can learn any linear threshold relation $\sum u_j x_j \geq 0$ or $\sum u_j x_j \geq \theta$. There is only one obstacle to implementing it directly on neuroids. This is that a coefficient v_j may become negative in the course of the algorithm, even if it is nonnegative in the linear threshold function that holds and is eventually learned. This violates the assumption that the values that the weights of any one edge can take have a predetermined fixed-sign (§5.2). The converse situation of a nonpositive fixed-sign weight going through a positive value is, of course, equally possible. Fortunately, it is easy to adapt the perceptron algorithm so that it works with fixed-sign weights, as is necessary here. The adaptation needed is that whenever an example would cause a weight to be updated so as to have a value with the disallowed sign, then this update to this weight is simply not carried out. It can be proved that this modified algorithm does learn any linear threshold function whose coefficients are consistent with the fixed-signs.[26] A second property that it is desirable to have if real neurons are being modeled, is that the threshold of the neuroid have some nonzero constant value throughout learning. The modified algorithm we describe in the Notes therefore works with a hypothesis of the form (9.2) with a fixed value $\theta^* > 0$, rather than 0, on the righthand side, that never changes in the course of the algorithm. While it may not be unrealistic to allow the threshold to vary in the course of a single interaction, we certainly need to avoid allowing the weights or threshold to grow uncontrollably in the course of a sequence of interactions, as the conventional version of the perceptron algorithm would allow.

Attractive alternative algorithms for linear threshold functions are also known. An interesting group of them resemble the perceptron algorithm in form, except for the significant difference that updates are multiplicative rather than additive (N. Littlestone 1988, N. Littlestone 1989a). As in the perceptron algorithm, updates occur only when a prediction is incorrect, and only those v_j are updated for which the current example has $e_j = 1$. One of these algorithms, called winnow2, assumes that every $u_j \geq 0$, and works with nonnegative v_j throughout. It initializes each v_j to be 1. For an appropriate constant $\alpha > 1$ it updates $v_j = \alpha v_j$ (rather than the perceptron's $v_j = v_j + 1$) in the case that the example is positive and the current hypothesis would predict negative. If the example is negative and the prediction positive then it updates $v_j = v_j/\alpha$ (rather than the perceptron's $v_j = v_j - 1$). The algorithm maintains a hypothesis $\sum v_j x_j \geq \theta$, keeping θ constant throughout. It can be analyzed and has particularly good behavior in cases where the number of relevant attributes, i.e. those x_j for which $u_j > 0$ in the true function, is small.[27] Littlestone has proved an upper bound on the number of times winnow2 predicts incorrectly for any, possibly infinite, sequence of examples, in terms of the parameters $k = \sum u_j, n, \theta$ and δ. The last parameter δ defines the distance by which $\sum u_j x_j$ separates the positive from the negative examples. In particular, δ $(0 < \delta \leq 1)$ and the u_j have to be such that for any positive example $\sum u_j x_j \geq 1$ and for any negative example $\sum u_j x_j \leq 1 - \delta$. Littlestone showed that if θ is chosen to equal n, then the number of mistakes made grows at most proportionally to $k \log n$ in any sequence of examples, provided δ can be regarded as a constant. A variant of it, called the balanced algorithm, can be adapted to deal with the general fixed-sign case and also maintains a fixed threshold, just like the adapted perceptron algorithm we describe in the Notes.

A problematic issue with any supervised algorithm for inductive learning is what to do if the function being learned cannot be expressed in terms of the knowledge representation assumed by the learning algorithm, or if learning fails for any other reason. We need algorithms to be *self-checking* in the sense that they can determine for themselves whether their current hypothesis is reliable.

The elimination algorithms for conjunctions and disjunctions described earlier in this chapter are both examples of local algorithms that are easily adapted to be self-checking. In the former case, if the examples are not consistent with any conjunction then all the variables will be eliminated after enough examples and all the incoming weights to the target node from the connecting circuit will be zero. The same will result for the algorithm for disjunctions once no new items are being added. If no item in the circuit is sufficient to guarantee that every example in which it is true is positive, then all of them will be eliminated from the candidate set. In other words, in both cases, the target nodes will be connected to the nodes of the hypothesized set $\{\tilde{x}_j\}$ only via edges of weight zero.

More complex local algorithms may need to be adapted more substantially if they are to become self-checking in a useful way. One generic approach is to have the $\underline{T}$ component of the mode of the target node keep track of some measures of reliability. In addition to having a numerical value $T^{(1)}$ for the threshold, one or more additional numerical measures would also be tracked. Examples of measures that are useful for evaluating the reliability of a hypothesis include the total number of misclassifications (and hence updates) performed to date in the course of learning at the current target, and the number of examples seen since the last one that was misclassified. These measures are easy enough to track by means of the neuroidal update function δ.

These performance measures are known to be useful indicators of reliability. For example, the main theoretical results about both the perceptron algorithm and Littlestone's winnow2 are upper bounds on the total number of mistakes made on any sequence of examples. Furthermore, it is also known that such bounds on the total number of mistakes imply good performance in the pac sense for algorithms such as the perceptron and winnow2 algorithms (D. Haussler 1988). Even better provable pac performance can be achieved if some modification is first made (N. Littlestone 1989b). An example of such a modification is to have the hypothesis that agreed with the longest sequence of examples retained as the most reliable one, even if it was subsequently contradicted.

Once such measures are recorded in the mode of the target neuroid they can be used in various ways. In the simplest case a node initially in state SL undergoes an initial learning phase. If and when its reliability measure exceeds a certain level, it proceeds to a new SLL state. In this new state the default threshold transition $T_i > w_i$ is allowed to operate. The node can then recognize the concept learned in the sense that it can undergo threshold firings triggered by the connecting circuit. In the training phase such firings are disallowed. Of course, if the reliability measure never reaches the required level, the node will be never considered to have learned.

This strategy can be adapted to allow the target node to be used for recognition while it is still learning. In this case a timing mechanism has to be employed to distinguish the situation of the target being fired by the connecting circuit being learned, as opposed to being fired by the circuit that computes the correct classification. Reliability measures are also essential in this case since if the examples and counter examples of the concept are not consistent with a linear threshold function, then the fact that the hypothesis generated contains no useful information has to be recognized.

Analysis in the pac model provides guaranteed performance bounds for arbitrary distributions or worlds. In practice, fewer examples or simpler algorithms often suffice. This can be explained sometimes by analyzing the relevant algorithms for particular distributions and arguing that the additional assumptions are not unreasonable. For example, it can be shown that for appropriate distributions the perceptron algorithm requires fewer examples to converge than the above analysis implies (E.B. Baum 1990). In some cases particularly simple algorithms suffice, ones that are provably highly resilient to errors (M. Golea and M. Marchand 1993).

Finally, we have to ask how these inductive learning algorithms fare in a hierarchical setting. The answer is that they are well suited. The complication created by hierarchies is that unintended lower level firings may accompany those occurring at the higher level. Fortunately all the algorithms we discussed in this chapter whether for conjunctions, disjunctions, or linear thresholds are robust to irrelevant attributes. In other words, given enough training examples, the hypothesis learned will not be perverted by such attributes. The only issue is the quantitative one that if these lower level firings are excessive, then the connecting circuits become large, waste space and may require many examples for reliable learning.

Chapter 10
Correlational Learning

10.1   An Algorithm

Humans appear to have an aptitude for noticing pairs of attributes that occur together with above average odds. We can easily identify acquaintances who are frequently late or streets that are often busy. In performing this task some attentional mechanism appears to mediate. It allows us to spot correlations only between those pairs of attributes that we have some reason to have noticed. Thus, even if it rained every Wednesday we may not observe this correlation unless this fact had some other special consequence for us. We hypothesize, however, that through learning experiences we can develop such an interest for *arbitrary* pairs of the attributes that we can recognize individually, and not just for those pairs that have some special semantic or other relationship. The problem we pose is, therefore, the following: How can a general correlation detecting capability, potentially able to spot correlations between arbitrary pairs of attributes, be feasible at all in a sparse network model such as the NTR we have defined?

Suppose that at an instant we have M items in the NTR with plenty of free space still unallocated. We hypothesize that there is an *attentional peripheral* that processes each input and, for the sake of argument, presents just two items to the NTR at any time. This is a special instance of Principle 3 that we shall discuss in Chapter 12 in connection with more general systems questions concerning the NTR. The assumption that the number of items here is two simplifies the presentation. We would find similar phenomena, however, if we had taken any other small constant number larger than two.

We first observe that, for any one random pair of the M attributes, it is extremely unlikely that the pair is presented by the attentional peripheral even once in a human lifetime for reasonable values of the parameters. Suppose that M exceeds 10^5 and that at any step each of the $M(M - 1)/2$ possible pairs occurs with the same probability (i.e. $2/(M^2 - M)$). Then for any fixed pair, say x and y, their conjunction occurs in any one step with probability less than about $p = 2 \times 10^{-10}$, since $M \geq 10^5$. But the number of inputs to the NTR in a human lifetime is probably best thought of as substantially less than $p^{-1} = \frac{1}{2} \times 10^{10}$ (since there are about 3×10^9 seconds in a century). We conclude that for any fixed pair of items x, y the expectation of ever having an input causing $\tilde{x}$ and $\tilde{y}$ to fire simultaneously is small, provided that x and y occur with the same frequency as the typical item, and they occur probabilistically independently of each other. Clearly, the probability of seeing such a conjunction more than once in a lifetime is even smaller.

This argument suggests that if any conjunction is witnessed two or more times, then it is reasonable to assume that the constituent items are statistically correlated. This fits in well with the scheme we described in Chapter 7 for unsupervised memorization. If a conjunction is seen once, then we allocate neuroids to it and regard it is representing the memory of some attributes that occurred together. If a conjunction is seen more than once, then we shall regard it additionally as the manifestation of an exceptional correlation. In that case we shall store in the neuroids representing the conjunction some measure of the frequency with which the conjunction has been observed. It seems impractical to maintain correlation information about every pair of items in memory, simply because the number of such pairs is too large. Hence some scheme such as ours that restricts the number of the pairs tracked to those that have been witnessed at least once seems essential.

We can regard correlational learning as taking place only at nodes initially in *available probabilistic* (AP) state. The algorithm that applies to this initial state is exactly like Algorithm 7.2 except that the state history AM $\rightarrow$ AM1 $\rightarrow$ UM is replaced by AP $\rightarrow$ AP1 $\rightarrow$ UP, and the history AM $\rightarrow$ AM1 $\rightarrow$ AM by AP $\rightarrow$ AP1 $\rightarrow$ AP. It has a $T^{(2)}$ component in $\underline{T}$ that measures the number of times the conjunction has been recognized. When state UP is reached for the first time $T^{(2)}$ is given value one. Each time the node subsequently undergoes threshold firings $T^{(2)}$ is increased by 1 (or perhaps by a decreasing amount as its value increases).

If we can afford to allocate new neuroids to any new conjunction of a pair of attributes experienced, then this simple approach is optimal. It can keep track of all the frequency information and hence no information is lost. Also this method can be extended to detect co-occurrences among sets of more than two elements by allocating storage to some pair first, and then to the conjunction of the item formed from this pair and the third element, etc.

On the other hand, the correlation detection capability of this algorithm is limited in other ways, since the algorithm assumes that each potential pair of attributes in a scene enjoys a separate presentation to the NTR. Thus if there are a hundred attributes in a scene we would need about 5,000 presentations. It would be interesting to determine whether humans do have correlation detection mechanisms that are too powerful to be explained by any algorithm such as ours in which only pairs or other small fixed size sets of attributes are presented to the NTR at any one time. Consider, for example, the following format for a psychological experiment. Suppose that there is a number, say 20, of lightbulbs. Each one, in each period of (say) one second, is on with probability p, and off with probability $1 - p$. The one second periods are synchronized for all the lightbulbs. Suppose also that the randomization controlling the lightbulbs is independent for all pairs, except for one unknown pair which is correlated (e.g. these two bulbs are on together with probability $\tilde{p}$ significantly higher than p^2). The problem here is that of determining the limits of human correlation detection in this setting in terms of the number of bulbs, the value of p, and the degree of correlation $\tilde{p}$ of the chosen pair. We note that even if we free ourselves from the constraints of our neuroidal model and are satisfied with detecting correlations on a von Neumann computer, there appear to be non-trivial lower limits on the number of computational steps required. For example, if we have m lightbulbs and choose $p = \frac{1}{2}$, no algorithm is known for detecting the pair with the highest correlation in a number of steps linear or close to linear in m. There is an obvious algorithm that checks out each potential pair in turn and, therefore, takes time quadratic in m. Some nontrivial improvements on this simple algorithm are known, but they do not adapt well to our neuroidal model.[28] It would be interesting to determine experimentally how far human performance can be pushed for this variety of correlation detection.

As pointed out in Chapter 3 learning correlations can be viewed as unsupervised inductive learning. It is induction in the sense that there is some element of inference that is not strictly deductive. It is statistical inference. If a conjunction is witnessed a large number of times, then it is inferred that the conjunction is indeed frequent in the world. It shares with unsupervised memorization, discussed in Chapter 7, the important aspect that it allocates its own storage. Consequently it has the same problem that, when an item $\tilde{x}$ fires, we do not want to waste space by allocating nodes that are adjacent to two members of $\tilde{x}$, which are clearly highly correlated. Hence breaking symmetry among a set of nodes, as Algorithm 7.2 does by timing, is also important here. Exactly as for unsupervised memorization we expect the peripherals to select by some attentional mechanism the pair of attributes to be presented, and to input each component of the pair at a distinct time.

10.2 Computing with Numerical Values

Since this chapter is concerned with information that has a numerical rather than Boolean nature, it is appropriate to address this broader issue here. The general position we have been taking is that for the functionalities considered in the previous chapters, the Boolean versions are the easier ones to conceptualize and will need to be solved first, even if more elaborate versions with real number parameters turn out to be more accurate. Of course, some of the functionalities, such as perceptron learning, adapt to the real number domain without modification.

It is believed that when a neuron in the cortex produces spikes at a significantly higher rate than its background rate, then some information is being conveyed. Furthermore, the frequency of the spikes is often associated with a real number that is some measure of the intensity level of the information. This immediately suggests a method for programming neuroids to process real valued data. Each number u that is processed will be represented as a sequence of firings of some neuroids. The frequency or rate will correspond to that number. The task of a neuroid will be to take such trains of spikes as inputs, and produce an output train of some desired, possibly different rate.

We shall give two illustrations of this. First, we shall show that a neuroid can compute a linear function of its inputs. Suppose that neuroid i is postsynaptic to neuroids $1, 2, \cdots, n$, and suppose that neuroid j $(1 \leq j \leq n)$ emits spikes at the average rate of u_j spikes per time unit where $0 \leq u_j \leq 1$. If $v_1, \cdots, v_n$ are real coefficients such that $u = \sum_{j=1}^{n} v_j u_j \leq 1$, then neuroid i will produce spikes at the average rate of u if it is defined as follows. For each j $(1 \leq j \leq n)$ let weight w_{ji} equal v_j, let $T_i := 0$ initially, and let δ be such that

$$\{T_i + w_i \geq 1\} \Rightarrow \{T_i := T_i + w_i - 1, f_i := 1\},$$
$$\{T_i + w_i < 1\} \Rightarrow \{T_i := T_i + w_i, f_i := 0\}.$$

Suppose that time t has elapsed and that neuroid j has fired u_j^* times in the first t time units. Then u_j approximates u_j^*/t and the desired value u approximates $(\sum v_j u_j^*)/t$. The transitions will make neuroid i fire every time the accumulated value of $\sum v_j u_j^* \approx tu$ increases by another unit. In other words it will fire at the approximate rate of u spikes per time unit, as desired.

As a second illustration, we observe that it is also easy to have a neuroid i recognize whether the weighted average rate of spikes u, as defined above, exceeds a certain rate $\bar{u}$. If we replace w_i in the righthand side of both transitions by $w_i - \bar{u}$, then neuroid i will produce spikes at the approximate rate of $u - \bar{u}$ provided that this is positive.

We note that in either application T_i is used to store a number rather than act as a threshold. Default threshold transitions are never invoked since the explicitly stated transitions cover all eventualities.

Chapter 11
Objects and Relational Expressions

11.1 Multiple Object Scenes

We regard a function of inputs as *propositional* if the function is best viewed as a predicate on the totality of the input. For example, when referring to a complex visual scene such predicates as "there is a green apple" or "there are three apples" can be viewed as propositions applying to the whole scene. Alternatively, if the input scene consists only of the image of an apple, then "green" is an appropriate propositional description. The representation of knowledge in the NTR can be viewed in the first instance as essentially propositional, each predicate being regarded as applying to the undivided totality of the input that is presented by the peripherals.

In order to have useful descriptions of the world, one appears to need more expressive representations than this. If a scene containing both a green apple and a yellow pear is presented to such a purely propositional system then the truth of each of the four predicates "green", "apple", "yellow", and "pear" may be recognized, but there is no way of representing the essential asymmetry among them, namely that the "green" is to be associated with the "apple" and the "yellow" with the "pear."

In classical predicate calculus logic this particular problem is overcome by introducing variables to represent the various *objects* that correspond to parts of the input. The case described here would be represented as:

$$\exists a \ \exists b \ \mathbf{green}(a) \ \wedge \ \mathbf{apple}(a) \ \wedge \ \mathbf{yellow}(b) \ \wedge \ \mathbf{pear}(b). \quad (11.1)$$

This expression states that there exist objects a and b with the required respective properties. Clearly there are great advantages of expressiveness in going beyond propositional representations. Unfortunately there are also very substantial computational costs. We know of no direct encoding of expressions of this kind with multiple objects that can be learned and manipulated satisfactorily in an NTR. Our solution will be to have a representation that is indirect, in the sense that it is purely propositional in terms of the NTR itself, but when interpreted through interactions with suitable peripherals can express multiple objects and more.

The fundamental technique that we use to enhance the expressive power of the NTR is *timing*. When expression (11.1) is to be memorized the peripherals will attend to the two objects a and b in turn. When attending to the first object the nodes representing green and apple will fire and their conjunction will be memorized. At a subsequent time the second object will be attended to and the conjunction of its two attributes memorized. Finally a *timed conjunction* (to be defined below) of the two conjunctions just learned will be learned. The neuroids representing this will fire in the future whenever the two lower level conjunctions fire at distinct times within a certain interval of time of each other. Thus the symmetry among the four predicates is broken by timing. During both learning and recognition any information presented to the NTR has to be broken up and time-stepped according to some appropriate *time schedule*. This task is executed under the control of the peripherals that implement the attentional mechanism.

In order to interpret information stored in the NTR in the correct nonpropositional way, we need to ascribe to the peripherals the abilities to segment scenes into objects, attend to each separately, and to schedule in time the necessary inputs to the NTR. We do not specify how the peripherals execute these tasks. The important point is that these tasks are not, in the first instance, random access tasks. Special purpose algorithms working from a visual image of the scene can be envisioned for them. The peripherals may use interactions with the NTR to help, but as long as no random access task (i.e. a task involving access to contents of the main memory other than through conventional access to the NTR) is ascribed to the peripherals, this approach is consistent with our methodology.

As noted in §4.5, there is ample psychological evidence suggesting that humans also use timing mechanisms for performing multi-object tasks. As the number of objects distinguished by predicate pairs increases in a scene, the more time it takes to process the scene. If less than sufficient time is allowed, then humans often perform the pairings incorrectly. For example, if a green apple and a yellow pear are presented for a short enough time, the four basic predicates may be correctly identified but some subjects will be confident that they have seen a yellow apple and a green pear.

11.2 Relations

We have seen that we need to go beyond a purely propositional representation if we are to represent multiple objects. Once we allow multiple objects, however, further opportunities for expressiveness present themselves. In the above example each predicate was *unary* in the sense that it applied to just one object at a time. With the possibility of naming several objects it becomes natural to allow predicates to apply and, in fact, to relate to several objects. Such predicates are called *relations*. For objects a and b typical relations may be **above**(a,b) to denote that object a is above object b, or **father**(a,b), to denote that a is the father of b. The order in which the objects are listed in the parentheses is considered to matter. Thus the semantics of **father**(a,b) is different from that of **father**(b,a). For certain particular relations the semantics may ensure that the order does not matter.

By means of relations one can write statements in the style of classical predicate calculus, such as:

$$\exists a \; \exists b \; \exists c \; \textbf{father}(a,b) \; \wedge \; \textbf{father}(b,c) \; \wedge \; \textbf{george}(a). \quad (11.2)$$

The literal interpretation of such a statement is that there exist objects called a, b, and c such that a is the father of b, b is the father of c and a has the property "george." Thus if one wished to express the fact that there is a grandfather called George, one could write such an expression.

In representing and computing with such relational expressions there is one pervasive problem called *variable binding*. This refers to the question of how correspondences among the objects are handled. The problem arises in at least two varieties, variable binding in *representation* and variable binding in *recognition*. To illustrate the former consider an expression of the form (11.2). How should a circuit in the NTR express the fact that the second argument in the first occurrence of **father** needs to refer to the same object as the first argument in the second occurrence? Since the NTR does not work with variable names, such as b, this conventional predicate calculus solution does not appear to work unless somehow implemented indirectly. Also, we cannot represent b by a neuroid representing a *particular* memorized item, since b is really a variable that may be satisfied by *any* person who is a father and at the same time the son of someone called George. Turning to the second aspect of the variable binding problem, that occurring in recognition, we consider now the problem of identifying whether (11.2) holds for a particular input. In this case the peripherals will have identified a number of objects in the scene and are left with the problem of determining whether there is an assignment of these objects to the object variables in the expression (11.2) that satisfies the expression. This latter task can be done, in principle, by trying all possibilities, but that would take time exponential in the number of object variables.[29]

Our approach here will be to define a class of expressions called L-*expressions* (for "labeled" expressions) that have less expressive power than the general expressions of predicate calculus, but offer the crucial advantage that they are computationally tractable on our NTR model. The basic idea is that, in addition to relations, these expressions also contain unary predicates describing each object that, in effect, label the object uniquely by some attributes. Thus we would make (11.2) into an L-expression if we conjoined it with the predicates

$$\textbf{old}(a) \;\wedge\; \textbf{middle-aged}(b) \;\wedge\; \textbf{young}(c), \qquad (11.3)$$

as long as no object simultaneously satisfied more than one of these additional predicates. It is an empirical question as to whether L-expressions, when learned hierarchically in the manner described here and starting from some plausible basis of preprogrammed functions, are indeed expressive enough to comprise a usable schema for describing relational aspects of the world. We are not suggesting that L-expressions can describe all useful relations. It is more than likely that several distinct such schemas are needed. Our purpose in discussing L-expressions in detail here is to demonstrate that at least some relational knowledge can be adequately processed in the neuroidal model.

We shall describe two positive computational results about L-expressions. The first deals with learning L-expressions containing only relations that are already implemented (i.e. preprogrammed in the system or previously learned). The second explains how new relations defined in terms of L-expressions can be learned.

We consider a relation to be *implemented* at a certain time if there are circuits in the NTR at that time that behave in an appropriate way when objects satisfying the relation are presented by the peripherals. In particular for a relation rel of, say, three arguments, there will be stored in the NTR three items rel^1, rel^2, rel^3, whose nodes can fire only after three objects, say, a, b, and c, satisfying the relation have been presented to the NTR. Furthermore, rel^i fires only if within a certain period after the identification of the three objects a, b, c, the ith among the three is presented again to the NTR, but by itself. In other words if $\text{rel}(a, b, c)$ holds and a, b, c are all presented together to the NTR within a certain period, then if any one of a, b, or c is presented separately within a certain period afterward, then the neuroids representing rel^1, rel^2, or rel^3 will fire at this subsequent presentation depending on which one of these three arguments is being presented.

It seems plausible to hypothesize that animals have the equivalent of some preprogrammed relations at birth. In order to be able to interpret the visual world, for example, an understanding of relationships such as "close to" or "bigger than" seems necessary. Our first result, therefore, describes how such L-expressions, involving only previously implemented relations, can be memorized.

In addition to this faculty, there seems to be a need also for some ability to acquire new relations. Hence, as our second result, we shall describe how a new relation, expressed as an L-expression in terms of preprogrammed or memorized relations, can be memorized. An example would be that of memorizing the notion of "grandfather" assuming that the relation "father" is already implemented. The aim is to set up circuits for the grandfather relation that will make the grandfather relation into an equal citizen with the father relation in future computations. We note that we are using the grandfather relation as an example simply because it is easy to discuss and not because it is among those for which there is compelling evidence that humans have a circuit.

For both of our results we use the mechanism of a *timed conjunction*, which we describe in the next section. As already mentioned, this is essentially like an ordinary conjunction, except that the condition for the target node to fire is that its arguments fire at distinct times within a given time interval, rather than simultaneously.

Our approach to relations is similar to the one used for multiple objects. The representation is essentially propositional, but becomes relational when interpreted by the peripherals through timing. Thus the items rel^1, rel^2, $\cdots$, rel^k, for a k-argument relation rel, are represented in the NTR exactly as propositional items. In the case that rel is preprogrammed the knowledge that these k items are somehow related is contained in the peripherals. For example, the relation $above$ being preprogrammed means that there exist items $above^1$ and $above^2$ in the NTR, that some appropriate peripheral can recognize the truth of a relation $above(a, b)$ when a suitable pair of objects a, b is presented, and that the peripherals can cause to fire $above^1$ or $above^2$ while simultaneously presenting a or b, as appropriate, to the NTR. Figure 11.1 illustrates this. (In the case that rel is a learned relation the corresponding relevant information is shared between the peripherals, which store the re-

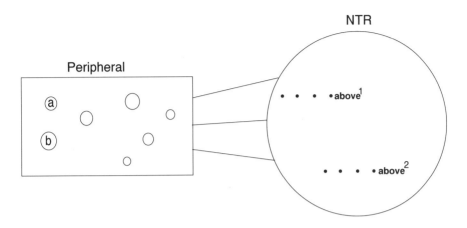

Figure 11.1. Schematic diagram of the implementation of the preprogrammed relation **above**. It is assumed that the peripheral can store images of a fixed number of objects, say seven, and can determine whether any of its preprogrammed relations hold for any subset of these objects. If it detects one such relation (e.g. **above**(a, b)) then it can schedule the inputting of a or b to the NTR and the firing of the **above**[1] and **above**[2] neuroids within the NTR, according to some schedule of its choice. In particular, it can resolve the binding problem by inputting a and firing **above**[1] simultaneously, and at a later time inputting b and firing **above**[2] simultaneously.

levant information about the constituent preprogrammed relations, and the NTR, which contains the newly learned circuits.)

11.3 Timed Conjunctions

A *timed conjunction* is a circuit in the NTR that has a similar role to that of regular conjunctions in Chapter 7. In order to make the target node fire, it is no longer sufficient, however, that the argument items fire once simultaneously. They must fire according to some other more complex time schedule. A timed conjunction will be denoted by $\wedge_t$ to distinguish it from an ordinary conjunction. The main application is for cases in which the arguments of the conjunction are items that are themselves conjunctions of various

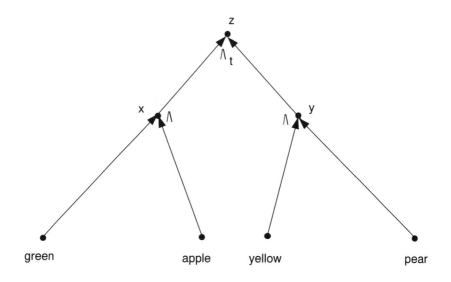

Figure 11.2. Schematic representation of a predicate involving two objects having the form of a timed conjunction of conjunctions.

sets of lower level items. An example would be the memorization or recognition of an expression such as (11.1). This would be done by a circuit of the form shown in Figure 11.2. This diagram and the ones to follow are schematic. For example, the replication of nodes representing any one item is omitted.

Here the top node is a timed conjunction and its two sons are ordinary conjunctions. Note that in a timed conjunction the target only fires if its arguments fire at distinct times within a certain interval.

In the memorization of the expression (11.1) by means of a timed conjunction as in this figure, the peripherals first identify the two objects as distinct, and memorize each separately as regular conjunctions to form circuits with targets $\tilde{x}$ and $\tilde{y}$ respectively. Subsequently the target nodes $\tilde{z}$ are programmed to act as timed conjunctions of x and y.

During recall the peripherals again identify the two objects. They present these objects to the NTR at distinct times so as to cause $\tilde{x}$ and $\tilde{y}$ to fire at appropriate distinct times within the interval required to make the timed conjunction target nodes fire in

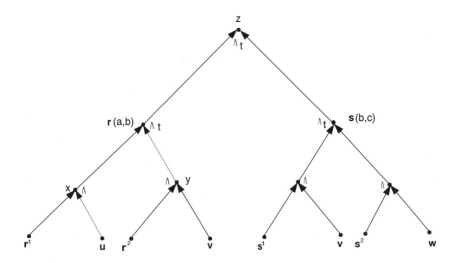

Figure 11.3. Schematic representation, using two levels of timed conjunctions, of the predicate (11.4).

turn.

We can also define *trees of timed conjunctions* of greater depth. These may be needed for expressing relational expressions, such as those of the form

$$\exists a \ \exists b \ \exists c \ \mathbf{r}(a,b) \wedge \mathbf{s}(b,c) \wedge \mathbf{u}(a) \wedge \mathbf{v}(b) \wedge \mathbf{w}(c). \quad (11.4)$$

This we will represent as a depth two tree of timed conjunctions as shown in Figure 11.3. The root timed conjunction is there to conjoin the two relations $\mathbf{r}(a,b)$ and $\mathbf{s}(b,c)$. Each of these two relations is itself expressed as a timed conjunction of ordinary conjunctions. Here $x = \mathbf{r}^1 \wedge \mathbf{u}$ expresses the fact that the first argument of $\mathbf{r}$ is the one with property $\mathbf{u}$, and $y = \mathbf{r}^2 \wedge \mathbf{v}$ the fact that the second argument of $\mathbf{r}$ is the one with property $\mathbf{v}$.

We now define the behavior of a timed conjunction more precisely. Suppose that τ_0 is some time period in terms of which the peripherals can synchronize and operate. Then a timed conjunction at depth one in a tree operates with period $\tau_1 = m\tau_0$ where m is the maximum number of arguments allowed in such a conjunction. In the examples above, all the timed conjunctions had two arguments. We envisage, however, that m may be some other small number also, such as three. The intention is that the target node will fire at some time within the interval $[j\tau_1, j\tau_1 + \tau_1]$ for some integer j, if its arguments fire within distinct subintervals $[j\tau_1 + k\tau_0, j\tau_1 + (k + 1)\tau_0]$, (i.e. for distinct values of k $(0 \le k \le m - 1)$). Thus if $j = 0, m = 3$ and if the timed conjunction in question does have three arguments represented by items x, y, z, then $\tilde{x}, \tilde{y}, \tilde{z}$ would each fire in a distinct one of the subintervals $[0, \tau_0], [\tau_0, 2\tau_0]$, and $[2\tau_0, 3\tau_0]$ for the target timed conjunction to fire. The definition allows the target to fire at the end of the last interval, at time $3\tau_0$ in this case. We envisage τ_0 as being rather larger than the smallest time unit of the NTR (i.e. corresponding to a macrounit rather than a microunit, in the terms discussed in §5.4).

There are several choices available for programming such a timed conjunction in the neuroidal model. We shall not go into the programming details here, but it should be clear that it can be done in a similar style to the neuroidal programs of the previous chapters.

For timed conjunctions at a greater depth in a tree of timed conjunctions, neuroids will need to be cognizant of larger intervals still. At depth two we need $\tau_2 = m\tau_1 = m^2\tau_0$. In general at depth i we need $\tau_i = m\tau_{i+1} = m^i\tau_0$. In other respects the implementation of a depth i timed conjunction in terms of one of depth $i - 1$ is similar to that of depth one.

11.4 Memorizing Expressions Containing Relations

A conventional propositional conjunctive expression is of the form

$$H = x_{i_1} \wedge x_{i_2} \wedge \cdots \wedge x_{i_l}$$

where each x_{i_k} is a predicate. In the propositional representation considered in previous chapters the firing of the neuroids corresponding to H and to each x_{i_k} depend on the truth of the corresponding predicate when applied to the totality of the current input to the NTR. Thus it would be more precise to introduce an object variable a to denote the totality of the scene throughout, and write the expression as

$$H(a) = x_{i_1}(a) \wedge x_{i_2}(a) \wedge \cdots \wedge x_{i_l}(a). \tag{11.5}$$

With such a terminology we can define an L-expression $H(a_1, \cdots, a_l)$ having k relations $rel_1, \cdots, rel_k$ and l object variables $a_1, \cdots, a_l$ as the conjunction of a "labelling conjunction" $H_j(a_j)$ for every object a_j ($1 \leq j \leq l$), and a relational statement $rel_i(a[i, 1], \cdots, a[i, m_i])$ for every relation rel_i ($1 \leq i \leq k$). Here rel_i has m_i arguments which we denote respectively by $a[i, 1], \cdots, a[i, m_i] \in \{a_1, \cdots, a_l\}$,. For the various values of i, the rel_i need not all be distinct relations. The same relation may occur with different sequences of object variables. In (11.2), for example, the relation **father** occurs twice. The $H_1, \cdots, H_l$ are each conjunctions of unary predicates applied to the respective a_i. Thus (11.3) can be interpreted as $H_1(a_1) \wedge H_2(a_2) \wedge H_3(a_3)$ where each H_i consists of one predicate. The resulting expression is an L-expression if $H_1, H_2, \cdots, H_l$ are *incomparable*. The syntactic definition of incomparability is that for any two distinct numbers j_1, j_2 ($1 \leq j_1, j_2 \leq l$) there is some predicate **u** in H_{j_1} that is missing from H_{j_2}. Then no H_{j_1} in the set is logically redundant in the sense that it is always true for an object whenever some other H_{j_2} is true. For L-expressions to be useful we will need in addition that the H_j be *semantically incomparable* with respect to H in the following sense: for any set of objects occurring in a "natural" scene that satisfies H, each object that satisfies some H_{j_1} will fail to satisfy any other H_{j_2}.

An example of an L-expression that conforms to this syntactic definition can be obtained, therefore, by conjoining the expressions (11.2) and (11.3). It expresses the grandfather relationship $gfs(a, b, c)$:

$$\textbf{father}(a, b) \wedge \textbf{father}(b, c) \wedge \textbf{old}(a)$$
$$\wedge \textbf{middle-aged}(b) \wedge \textbf{young}(c). \tag{11.6}$$

We note that some of the predicates which we describe here as

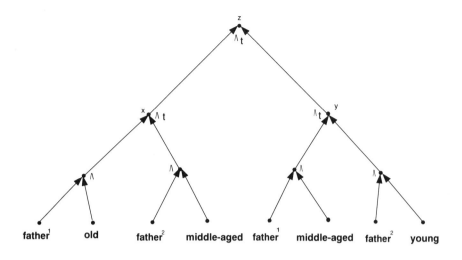

Figure 11.4. Schematic representation using two levels of timed conjunctions of the grandfather relationship defined in expression (11.6).

unary, such as **old,** may be more accurately viewed as having in reality two arguments, the particular object they apply to as well as the totality of the scene. In other words the totality of the scene or context may be useful as an extra argument in all the predicates and relations. In this example the truth value of old may not be determined by a strict numerical criterion of age, but some, possibly more complex condition on the totality of the input, in this case perhaps the age relative to that of others in a group photograph. Clearly, however, we need to avoid the circularity of having a relation defined in terms of predicates that already express the same meaning.

Our claim is that L-expressions are expressive enough to be useful, and, at the same time, both varieties of the variable binding problem can be solved for them in the NTR model. The computational tractability of L-expressions is based on the fact that they can be represented by timed conjunctions of depth two. The expression (11.6) can be represented, for example, as shown in Figure 11.4.

Here $\tilde{x}$ is a timed conjunction recognizing $\mathbf{father}(a, b) \wedge \mathbf{old}(a)$ $\wedge$ **middle-aged**(b) and $\tilde{y}$ is a timed conjunction recognizing father(b, c) $\wedge$ **middle-aged**(b) $\wedge$ **young**(c). The identity of the second argument of x and the first argument of y is ensured by the fact that they have the same precondition to fire, namely the predicate **middle-aged**. Also, by incomparability, when this condition holds, and the nodes representing **middle-aged** fire, then the conditions that label each of the other objects will fail to hold.

When memorizing such an L-expression, with the constituent relations already preprogrammed or previously memorized, the peripherals need first to identify the objects in the scene and the relevant relations among them. Note that we are assuming that the peripherals can evaluate the truth of any of the preprogrammed relations for any subset of the objects depicted in the relevant peripheral. Once the three objects a, b, c have been selected, and the truth of the relations father(a, b) and father(b, c) confirmed, the peripherals focus on the four argument occurrences in turn. First a and father^1 are input, causing all the predicates that hold for a to fire at the same time. A conjunction of father^1 and all these unary predicates will be memorized at this time. Subsequently father^2 and b are input in similar fashion. After that the timed conjunction $\tilde{x}$ is memorized, and this is achieved by ensuring that the two constituent conjunctions just memorized are made to fire according to the appropriate schedule. In similar fashion $\tilde{y}$ is memorized. Finally by inputting again the appropriate objects a, b, c and making father^1 and father^2 fire, all according to an appropriate schedule, the target timed conjunction $\tilde{z}$ will be memorized.

When recognizing this L-expression, the peripherals again have to identify the three objects a, b, c and the validity of the two relations father(a, b) and father(b, c). They will need to input this information to the NTR according to a time schedule similar to that used for memorization. We note that for such timed conjunctions of depth two the only knowledge required to determine an appropriate schedule is the number of relations and the numbers of their arguments. Since there is no requirement on the ordering in time of the firing of the constituents of a timed conjunction, widely differing time schedules may be employed on different occasions when recognizing the same relation.

11.5   Memorizing New Relations

In the previous sections we assumed that any relation occurring in the expression to be memorized is already implemented (i.e. preprogrammed or previously memorized). As outlined previously, a relation $\mathbf{rel}(a_1, a_2, a_3)$ is considered implemented if there are items $\mathbf{rel}^1, \mathbf{rel}^2, \mathbf{rel}^3$ in the NTR such that (1) they never fire unless three objects satisfying the relation $\mathbf{rel}$ have been identified, and (2) $\mathbf{rel}^i$ does fire if within a certain period following the identification of the three objects, the object a_i that fits as the ith argument of $\mathbf{rel}$ is input by itself.

When memorizing a new relation we have to set up a circuit having just these properties. Suppose we want to memorize the relation $\mathbf{gf}(a, c)$ defined as:

$$\exists b \ \mathbf{father}(a, b) \ \wedge \ \mathbf{father}(b, c) \ \wedge \ \mathbf{old}(a)$$
$$\wedge \ \mathbf{middle\text{-}aged}(b) \ \wedge \ \mathbf{young}(c). \qquad (11.7)$$

The task of memorizing this is more onerous than that of memorizing the predicate $\mathbf{gfs}(a, b, c)$ of (11.6). In the case of gfs we are memorizing just one item, while in the case of gf we are memorizing two items $\mathbf{gf}^1$ and $\mathbf{gf}^2$. The difference is that in the latter case we wish to be able to use gf as a constituent relation in arbitrary L-expressions at later times, with the same flexibility that preprogrammed relations allow.

The circuit that needs to be set up is shown in Figure 11.5. Here the circuit up to node z acts exactly as in Figure 11.4. We will have z continue to fire for a longer period than before, once it has been caused to fire. If, within this period, either the object satisfying $\mathbf{old}(i.e.\ a)$ or the one satisfying $\mathbf{young}(i.e.\ c)$ is input, then the corresponding one of $\mathbf{gf}^1$ or $\mathbf{gf}^2$ will fire. This is exactly what is required for a relation to be implemented. Memorizing such a circuit requires a time schedule only slightly more complicated than that for memorizing an L-expression.

The conclusion, therefore, is that the mechanism used for memorizing L-expressions can be used also for the more complex task of memorizing a new relation, provided that the latter is expressible as an L-expression in terms of previously implemented relations.

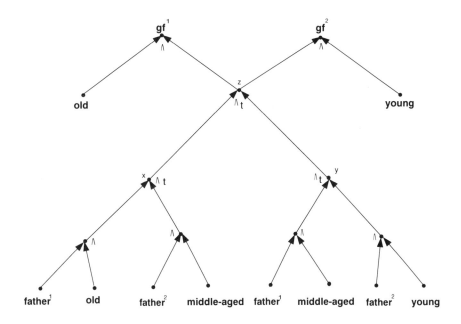

Figure 11.5. Schematic representation of the grandfather relationship components gf^1 and gf^2 of the relation $gf(a, c)$ defined in expression (11.7).

11.6   Discussion

We see that establishing circuits for expressing relational information appears to be substantially more involved than is the case for propositional information. Much more is required of the peripherals both during memorization as well as recognition. It is not clear what degree of intricacy in the mechanisms required should be considered acceptable. There is considerable evidence that in the brain the task of laying down long term memory is a complex one involving little understood interactions between the hippocampus and the cortex that may sometimes take at least several weeks to complete. Hence we should expect that at least some of the basic algorithms involving the NTR would be as complex as the L-expression algorithms we have described. On the other hand, the difficulties of dealing with relations suggest to us that relational information is memorized as sparingly as possible, and all manner of tricks might be used to compensate for this shortcoming.

It is important to note that the algorithms in this chapter dealt with what is the most onerous case, where relational information has to be recognized in a single interaction with the NTR. In Chapter 13 we shall discuss the issue of multiple interactions in which the information obtained from the NTR by the peripherals in one interaction may influence the next interaction with the NTR. Consider the human ability to recognize the outline map of a country, such as France, an example we discussed earlier. We appear to be able to make a correct positive identification, even if the outline is partially inaccurate or the map has on it additional data that we recognize as erroneous. This suggests that our recognition algorithm does not contain all our knowledge about France. It is merely a "rule of thumb" procedure or "reflex" that we have acquired and is, presumably, reliable in making us think of France when that is appropriate. Once recognition at this level has taken place, any amount of additional information may be evoked as a consequence. In this instance of a map, this may mean the names of a certain number of major cities and their relative positions. This additional information may involve a substantial number of relations. It is not essential, however, to have this relational information as an integral part of the basic concept of a map of France that is invoked at the first level of recognition. It can be accessed by association whenever recognition at this first level occurs.

The algorithm for recognizing the grandfather relation should be considered in a similar light. First, it is possible that in practice humans learn the explicit definition of a grandfather, in a form similar to (11.2), and reason with it, formally (by means of "Turing reasoning" in the sense of §13.2) in the course of a sequence of interactions with the NTR. The relation "cousin four times removed" is an example of a relation that is almost certainly in that category. It seems implausible that we need a precompiled circuit able to recognize this relationship in one act of circuit evaluation. The complications described in this chapter are, of course, not relevant to relations for which precompiled circuits are not needed. Furthermore, even if the NTR does contain a circuit equivalent to Figure 11.5 for grandfather, this circuit needs to be used only when identifying for the first time whether this relationship holds for a set of individuals. Once the identification has been made that

George is the grandfather of Joe, this fact can be stored using conventional propositional representations, without the need for timed conjunctions. For example, identifying George with the conjunction of Joe and gf^1 will have this effect. Thus the strategy of choice for the NTR is to store information in purely propositional form whenever possible.

In spite of these disclaimers, we believe that some account, such as ours, is needed to explain how fast recognizers for concepts involving relational information can be supported at all in the NTR. Suppose, for example, that on seeing a group photograph one instantly recognizes it as a family group and identifies the grandfather in it without any conscious reasoning process being used. One possible mechanism for performing this identification is the evaluation of an appropriate L-expression. The facts that the person we identify as the grandfather is, in fact, old, and that there are other people on the photograph who are young and middle-aged, respectively, appear to facilitate our recognition process. The labeling of the objects by distinguishing predicates, which is what L-expressions contain, seems relevant and useful.

As mentioned earlier, it is probable that not one, but several schemas or tricks are needed to cope with the full spectrum of relations humans need. Clearly relations in space, such as **above**, are vital to us, and we may have special purpose help in handling them in our vision system. Also, it is plausible that we make full use of any special facility that we may have with spatial relationships in other areas also, by using analogies that map otherwise unrelated problems to spatial ones. In some instances we may imagine the family relationships we discussed by visualizing the grandfather, father, and son as arranged from top to bottom spatially according to age, and this may help us reason instantly about them. Our discussion of L-expressions is intended to show merely that relations in some generality can be accommodated within the neuroidal model. In many instances other methods may be more appropriate or computationally simpler. Consider, for example, relations such as $\mathbf{samecolor}(a, b)$ or $\mathbf{sibling}(a, b)$ that are symmetric in the sense that the order of the arguments does not effect the semantics. L-expressions are not well suited to such relations. Alternative mechanisms may be preferable when reasoning about them. For example, one may hypothesize a mechanism that in any situation regards the equivalence class of all objects that are brothers, or of the same color, as one entity.

Chapter 12
Systems Questions

12.1 Introduction

In previous chapters we demonstrated how any one instance of any one of several different functionalities can be realized in an NTR. When we wish to realize all these functionalities together and to invoke them not just on one occasion but in arbitrary sequences, then significant additional problems arise. These *systems* problems fall broadly into two categories, depending on whether they relate essentially only to the internal working of the NTR or whether they concern directly the particulars of the interactions with the peripherals.

In the first category are the problems of ensuring that the functionalities have compatible implementations that allow their invocations to be *cumulative*. We need to explain how, even after a long sequence of updates, using each of the several algorithms that implement the various functionalities, the NTR is still in a suitable condition to execute further invocations without unduly disturbing the effects of the earlier ones. This cumulative feature needs both qualification and quantification since, clearly, if knowledge is constantly added and never deleted, then the NTR will run out of storage space eventually. This particular problem could be overcome, in principle, by allowing a "forgetting" mechanism. We have not included one, mainly because we do not know of a natural criterion of what should be forgotten. A further reason is that the numerical estimates discussed earlier, of neurons and of lifetime limits for humans, suggest that it may not be strictly necessary. We note that in line with our general approach, we wish to solve all the systems problems in the NTR by mechanisms that can be implemented within the model that we have defined, and do not require additional mechanisms, such as global operations. It is quite possible, of course, that global operations realized by chemical action, for example, do play important roles in the brain.

Systems problems falling in the second category arise necessarily in any neuroidal system where peripherals mediate between the NTR and the outside world. Since the outside world may be arbitrarily complicated, a first task of the peripherals is to restrict the amount of information presented to the NTR at any instant. A second role is to perform this restriction in a useful way. In more psychological terms, what the peripherals need to do is to solve the problem of focus or attention, which is concerned with restricting the incoming information at any time to that which describes particular objects or semantic units in the world.

12.2   General Organizational Principles

In order to make the various mechanisms described have the desired effect in the neuroidal system we find that the following three general principles appear to help. We shall turn in the next section to the issues of how the particular mechanisms described for the NTR can be made compatible.

Principle 1: The mechanisms that set up connections in the NTR in one direction (i.e. LINK, JOIN) have analogues that can set up connections in a reverse direction. For example, when an image in a low level vision peripheral can cause the neuroids of a high level item in the NTR to fire, mechanisms for the inverse causation of firings are also available. In particular, the firing of the high level item will cause firings in the peripheral, corresponding to the appropriate image.

Principle 2: The nodes in the NTR that are firing can detect for themselves whether they are the highest levels of nodes firing at that time. When a node fires and detects itself as being at the highest level of firing, it is said to be *charged*.

Principle 3: The peripherals, in conjunction with the NTR, implement an attentional mechanism that at any instant imposes the following limitations on the NTR: (i) it limits the input to the attributes of one or a small number of objects in the scene observed, and (ii) it limits the number of nodes charged to a fixed small number M_e of items. The latter limitation implies that for any object in the scene only a bounded number of attributes can be given the highest level of attention.

Principle 1 seems essential if we desire that changes in the NTR be able to effect action in the outside world. If we wish that the firing of higher level neuroids in the NTR result in changes in the peripherals, then connections realizing the associations in this reverse direction need to be in place.

Implementing Principle 1 poses no fundamental problem. Every time a new forward connection is implemented by a LINK or JOIN operation in the NTR, (whether by a simple direct vicinal algorithm as in Chapters 7 and 8, or in some more involved way as in Chapter 14) we can follow it by a LINK operation in the reverse direction, using new relay nodes. Although, as we claimed, there may be no fundamental impediments to doing this, the details of how it should be done to obtain the desired behavior may be complicated. It may turn out, for example, that in certain human memory disorders the difficulty with adding near memories is not with the allocation of new storage, but with these reverse associations which are necessary for recall.

In one method for realizing Principle 2 it will be useful for relay nodes that realize reverse connections to be in a state that distinguishes them as "reverse relay" nodes, both before and after they have been allocated. This can be achieved by, for example, having relay nodes of the two kinds initialized to different states. For our applications it is sufficient, in fact, that on each reverse chain at least one relay node be distinguished as "reverse relay."

Principle 2 can be implemented in various ways, among which the following is one. Suppose that Principle 1 is implemented and in each reverse connection at least one node is in "reverse relay" state. Suppose also that once such a node is fired it emits a train of signals in a time pattern unique to reverse relay nodes. Suppose further that in a cascade of firings, once a non-relay node fires it remains in a "pseudo-refractory" condition until stability is reached. In this condition the node keeps firing, independent of its neighbors, but its state may otherwise depend on its neighbors. We can program this pseudo-refractory condition so as to be able to detect the signature of any reverse relay node from which signals are coming via relay nodes. The result will be that every non-relay node in the cascade that is not at the highest level will receive a signal train from a reverse relay node, but those at the highest level will not. Hence nodes in the pseudo-refractory condition that do not receive the reverse relay signal can rightfully identify themselves as charged.

A different approach would be to dispense with reverse relay nodes having distinguishing signature patterns of firings, and instead have weights on reverse edges that differ sufficiently in magnitude from those on the forward edges that nodes can distinguish the direction from which a signal is coming. Then nodes that do not receive reverse signals can again recognize themselves as charged.

One important consequence of Principle 2 is that it allows for supervised learning at target nodes that have been allocated hierarchically. For example, if the recognizer for the sound of the word "dinosaur" is itself a hierarchy, the desired result of learning the meaning of the word is that, on seeing examples of a dinosaur, the highest level nodes of the sound hierarchy should fire, and not lower level ones such as those corresponding to "dino." In order to achieve this we need that the highest level nodes in the sound hierarchy be able to distinguish themselves by becoming charged and going into a different state than the others. They can then correctly learn in supervised mode without any undesirable side-effects.

Principle 3 is concerned with attentional mechanisms, among which we distinguish two varieties. The first provides a way for the system to identify parts of scenes that are integral objects, and to input to the NTR attributes of such objects as well as relationships among a small number of them. In a visual scene such an object may be a group of people, a person, a face, or a nose. The choice among these will be determined via interactions between the peripherals and the NTR. Some such interaction appears also to be useful for isolating the attributes of objects, since knowledge that is not preprogrammed in the peripherals resides in the NTR. Also, some version of Principle 2 seems necessary if we are to distinguish one level of the knowledge representation hierarchy (e.g. a face) from the others (e.g. a nose). The peripherals will have some further special purpose functionality in addition. For example, a low level vision peripheral may have capabilities for detecting boundaries and color information, that do not require interaction with the NTR.

Principle 3 also states that a second type of attentional mechanism is needed, that restricts at any time the total number of neuroids that are effectively in charged condition. This is important for several technical reasons. First, if we have l items firing and we allocate new storage by unsupervised memorization as in Chapter 7, then we are allocating storage for all the $l(l-1)/2$ different pairs. If the value of l is not restricted, then storage might be used up much too fast. If, on the other hand, we require an item to be charged in order to take part in any such allocation, then by restricting the number of charged items to, say, $l = 5$, we keep good control of memory utilization. A second context in which Principle 3 is beneficial, is in implementations, such as Algorithm 8.1, where relay nodes are shared among many target items. In order to avoid target items being fired by the firing of spurious combinations of nodes, it greatly helps if the number of nodes firing at any time is restricted. The same effect can be achieved by restricting the number of charged nodes and ensuring that only these have influence. Of course, we need to adapt the corresponding earlier algorithms so that they are influenced only by charged nodes, which distinguish themselves from the others somehow by, for example, their pattern of firing. More generally, our previous algorithms for memorizing conjunctions in Chapter 8 and L-expressions in Chapter 11, should be reinterpreted to mean that the predicates being conjoined are not all the items true for the input scene, but only those that are being attended to, in the sense that the neuroids representing them have become charged.

To implement this second kind of attentional mechanism one appears to need some process that involves the NTR interacting with some kind of imagery peripheral. This interactive process would stabilize when a small enough set of items is charged in the NTR, a set that corresponds to a consistent image. Such a mechanism necessarily has to resolve among competing sets of predicates. When we view a picture that has two alternative interpretations, we clearly need to use some such resolving mechanism. In more frequent situations a picture has just one dominant interpretation, and the alternatives are not serious competitors. One can, therefore, hypothesize some iterative improvement strategy that finds a good consistent interpretation. This may be most aptly viewed as a continuous rather than discrete process. For example, the lengths of the spike trains may play a quantitative part. Although the choice of this resolving strategy is crucial in determining the behavior of a neuroidal system, it is enough here to hypothesize that one exists. Further, we hypothesize that it is the ability to learn from situations where the wrong conclusion is reached that makes it effective. In any instance in which the resolving strategy gives an unsatisfactory answer, some learning in the NTR will take place that will give higher weight to the correct solution when a similar instance is presented in the future. Clearly we cannot attempt to specify the particulars of the resolving strategy without specifying the particulars of the peripherals. As we have argued earlier, we omit giving these particulars and this omission is consistent with our declared methodology. Substantial research efforts have been expended, however, in studying such iterative optimization processes in other neural network models, and the phenomena found there may translate to our framework.

Finally, we need to discuss whether it is plausible that the human brain implements the three principles outlined above.

As a preliminary we need to repeat that there may be several alternative correspondences between real neurons and spikes, on the one hand, with neuroids and their firings, on the other. A neuroid may model either one neuron or a group of them. A firing may correspond to a single spike or a train of them. On this second point we take the view that the recognition of Boolean items in neurons is communicated by trains of spikes. In some of our mechanisms, as in simulating LINK and JOIN on sparse random graphs in Chapter 14, we do use such trains. Even in cases where the high level algorithms do not require them, it would appear that an implementation in terms of spike trains may be desirable. The length or frequency of the train may provide a numerical value, beyond the Boolean bit, that adds richness to the computations, in dimensions that we have not modeled here. This general picture of information being conveyed in trains of spikes is consistent with experimental evidence. For example, neurons in monkeys that have been identified as responding to the sight of faces, transmit long bursts of spikes (e.g. 100 in half a second) when the monkey is presented with a view of a face. In the absence of any such stimulus the same neuron produces spikes apparently randomly, and at a lower rate (e.g. 5 per second.) These single random spikes, according to this general picture, provide background noise to the more purposeful events that are modeled by vicinal algorithms.

In low level vision areas of the cortex it is found that a complex scene presented to an animal may cause a substantial percentage of the neurons to fire at above the resting rate. This is consistent with our hierarchical representation of knowledge. Presumably a substantial fraction of these neurons correspond to low level items that are each made to fire by a significant fraction of natural scenes. In visual scenes such items may correspond to circular patches, corners, etc. It may be that in low level sensory areas of the cortex one needs larger numbers of less dominant responses, while in higher areas the few neurons that correspond to each high level item, fire in more dominant bursts on the rare occasions that they are called on to do their jobs.

For Principle 1 there is some corroborating evidence that has been referred to earlier. Neuroanatomists have investigated in great detail which areas of the cortex have connections to which others, and which have connections to subcortical areas of the brain. It has been found, almost invariably, that whenever there are connections in one direction between one area and another, there are also connections in the reverse direction.

We conclude by noting again that we are not claiming that there is convincing evidence that the algorithms and mechanisms we describe here are those that are implemented in the human cortex. We are claiming only that we have a fairly detailed model that is demonstrably able to implement some interesting cognitive functions, and that its nature and mechanisms are not inconsistent with current knowledge of the brain. In the light of new knowledge the model may need to be refined, adapted, or changed. But the validity of any alternative model will have to be demonstrated, in the same spirit as here, by exhibiting concrete algorithms that demonstrate that, at least in principle, it can support interesting cognitive functions.

12.3   Compatibility of Mechanisms

In earlier chapters we described neuroidal algorithms for a variety of tasks with the suggestion that all of them could be supported simultaneously in a single system. We refrained, however, from detailing a single "megasystem" that incorporates them all, both because that would have introduced excessive complications, but also because we do not wish to discourage the exploration of alternative algorithms for any one of them. The question of whether the algorithms described contain major contradictions is, of course, a significant one. We believe that they do not. We shall illustrate here how this issue might be considered more systematically by considering just one instance of it, namely the co-existence of unsupervised memorization with inductive learning.

Consider the process in which the spelled word "dinosaur" is first memorized, and subsequently a recognition algorithm for this concept in terms of other categories, such as "extinct" or "large", is learned by induction. Figure 12.1 shows a schematic diagram of the circuit that results from this process when the algorithms of Chapters 7 and 9 are invoked. The right hand side of the diagram illustrates the hierarchy of nodes allocated by unsupervised memorization by the fragments of the word. The choice of fragments may, of course, depend on various factors such as whether the input of the spelled word was visual or auditory. The final result of this part of the process is to allocate to the set of nodes at the top the word "dinosaur" and to have these nodes in state

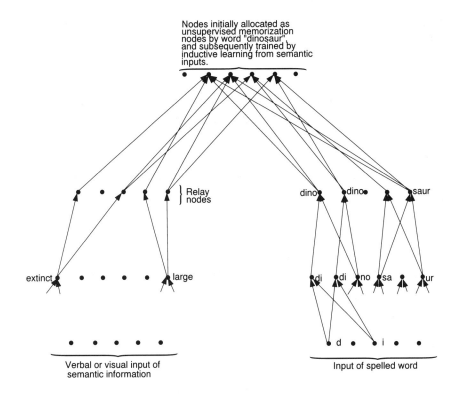

Figure 12.1. Schematic illustration of a subcircuit. The nodes at the top were initially allocated by unsupervised memorization following the input of the spelled word "dinosaur." Subsequent to this allocation, these same nodes induced some information as a result of a number of further inputs, in each of which the spelled word "dinosaur" was presented in conjunction with additional semantic information about this animal.

UM. The second part of the process, that of inductive learning, is responsible for setting up the circuit shown in the left hand part of the diagram. First, communication between the relevant semantic nodes, such as those representing "extinct", and the target nodes is established by means of relay nodes. Subsequently the weights in this part of the circuit are adjusted so that it will subsequently evaluate to the function that has been induced.

In order to verify that the algorithms involved can be implemented compatibly, we need to examine the assumed preconditions of, and the transformations resulting from each one. In the case at hand, prior to the unsupervised memorization process the target nodes are in state AM and have incoming weights all equal to one. After these nodes have been allocated their state will be UM and the weights on edges not involved in transmitting information from the spelled inputs will be zero. The remaining weights and thresholds will be such that a future presentation of the word "dinosaur" will make these target nodes undergo threshold firings. Suppose now that these UM nodes have a capability to learn conjunctions inductively. (We could equally assume here disjunctions or linear threshold functions. We could even assume that each node attempts first to learn one of these, say conjunctions, and if that fails it then tries another class.) The inductive learning algorithm will make nonzero some of the zero weights, namely those on edges coming from the relevant semantic items via the relay nodes. A circuit recognizing the induced function will be created. The weights and thresholds of the target nodes will then be such that at later times they will fire whenever either the word or the semantic description is presented.

The issue of compatibility has to be addressed explicitly when one attempts to model significant psychological phenomena that involve more than one functionality. As an illustration consider how the circuit shown in Figure 12.1 might be used to model the phenomenon of *priming* that we discussed in §4.5. This phenomenon is concerned with a subject being exposed to information that is not consciously recollected later, but that measurably changes the subject's performance in a related task at a later time. If one has explicit hypotheses about the circuits of memory, then one can make explicit theories about the nature of priming. As an illustration consider the theory that when an input has caused a subcircuit to undergo threshold firing, then the edges whose weights contributed to the firing of some node will increase for some temporary duration. For any particular set of circuits such a theory will make predictions and these can be tested by experimentation. For the simple circuit shown in Figure 12.1, for example, some predictions can be read off immediately. An example of such a simple prediction is that exposure to a spelled word will produce different priming effects from those produced by exposure to a picture, since these two conditions will change different sets of weights.

In particular, performance will be increased most for those inputs that have the same modality as the original priming, an effect that is consistent with current psychological evidence. There is a large body of experimental results about priming already available (D.L. Schacter 1992, J. Bowers and D.L Schacter, 1994) and this may be one of the areas of psychology where detailed neuroidal modeling may be already viable and fruitful.

Chapter 13
Reasoning

13.1   Introduction

Adaptability and flexibility are traits that are particularly associated with humans, who can often react to situations not previously experienced with inexplicable speed and competence. These traits have proved particularly difficult to emulate in machines. There have been many attempts to endow computers with large amounts of knowledge, but for large systems the result is usually brittle. The systems do not work competently in situations not foreseen by the programmer. This failure contrasts with the often impressive performance even of certain nonhuman species. The following experiment has been reported (D. Premack 1983): Two widely separated deep containers in a room or field are pointed out to a chimpanzee. An apple is then placed in one and a banana in the other before the animal is temporarily removed from the scene. The animal is then brought back, at which time it sees the experimenter eat a fruit, either an apple or a banana. Does the chimpanzee behave sensibly by moving toward the bin containing the remaining fruit? The answer reported is yes. Performance on this task was found to be comparable to four and a half year old humans.

What view should we take of this kind of behavior? Aristotle claimed that all knowledge is derived either through what he called induction or through deductive reasoning, to which he referred to as syllogism. Our treatment follows essentially the same distinction, but perhaps with a different emphasis. While his writings tend to detail and highlight reasoning over induction, the balance here is tilted in the opposite direction.

A summary of our general viewpoint is the following: Much knowledge and most skills are acquired inductively. These acquisitions take the form of a large number of short programs or reflexes, which are useful for coping with inputs from the world or for resolving among other reflexes. These reflexes derive their validity only from the fact that they work in the complex world and can be justified intellectually in the framework of pac learning. The availability of large numbers of inductively learned reflexes is not quite enough, however, to provide an explanation of the full range of human intellectual responses, or even of the chimpanzee behavior described above. In new situations we appear to be able to put together information from a well chosen subset of these reflexes that we had no reason to consider together before, and to arrive at a new course of action. Each time we hear and understand a sentence containing some new or surprising information we must be doing something like this. It is internal computations of this kind, that pull together a novel combination of pieces of information already known, that we consider here to characterize *reasoning*. Aristotle's notion of syllogism may be considered perhaps as a special case of a more general class of phenomena.

Formalizations of reasoning are often at odds with formalizations of inductive learning. One central problem is that the notion of consistency is at the heart of most formal systems of reasoning. For example, in some systems each proposition is either true or false, and the deductions that can be made have to yield consistent truth values for all the propositions that can be expressed within the system. In pac learning reflexes relate to the world in a statistical manner. More importantly, it is acknowledged that the statistics of the world are typically complex and the system should be able to cope with the world effectively without needing to know much about these complex statistics. The traditional insistence on consistency is somewhat at odds with this view of knowledge, though some reconciliation is nevertheless possible.[30]

The view of reasoning that emerges from the neuroidal model is the following. Knowledge can be added to the NTR both by memorization as well as by induction, as the algorithms described in the previous chapters suggest. There are additional processes, which we call reasoning, by which new conclusions can be reached based on information already stored in the NTR. These processes may need the use of the peripherals, not only to schedule firings, but also as temporary repositories for the information that is newly brought together in a deduction. The peripherals that take part we shall call *imagery* peripherals. This choice of words is intended to suggest that visual metaphors and concrete examples, rather than the abstractions of mathematical logic, may have the more central role in human reasoning.

The aspect of reasoning that we are particularly concerned with here is so called *commonsense reasoning*. This is the process that humans use to cope with the mundane but complex aspects of the world in evaluating everyday situations. It is reasoning that is generally done subconsciously. It is perhaps precisely because we have no awareness of these processes that it has proved so difficult to simulate them in machines. No one has yet made a home cleaning robot that can execute its task with reasonable flexibility and commonsense.

We believe that the subconscious processes that manifest themselves as commonsense reasoning constitute fundamental parts of the substrate of cognition. The more complex reasoning tasks that are typified by puzzle solving, and which humans perform consciously, may use them as building blocks. We shall not pursue here the phenomena of higher level reasoning, aspects of which have been investigated extensively by psychologists (K.J.W. Craik 1943, P.N. Johnson-Laird 1983). Some activities that are conventionally associated with higher level reasoning, such as chess playing, have proved easier to mechanize than commonsense reasoning, though not by methods that attempt to emulate the psychological processes.

13.2   Reflex Reasoning

We shall distinguish among three kinds of reasoning activity in terms of the computational demands they make on the neuroidal system. They are of increasing power and make increasing demands on the peripherals, but all work with an NTR that is constrained by the neuroidal model described in earlier chapters.

The most basic of the three kinds of reasoning responses consists of the peripherals presenting some information to the NTR and the NTR undergoing a single cascade of threshold firings as a result. We shall call this *circuit evaluation* or *simple reflex* reasoning.[31] The computations involved are essentially the same as those used in recognizing a memorized item or an example of an inductively learned concept. We shall argue, however, that these computations can also be viewed as supporting the kind of simple reasoning that humans appear to perform particularly fast. Within the category of simple reflex reasoning we shall allow processes in which the main act of circuit evaluation is preceded by some interactions between the NTR and the peripherals that perform such auxiliary tasks as scene segmentation. We shall further include tasks that may involve several circuit evaluations, each prompted by the attention mechanism focusing on distinct parts of the input successively. The main criterion that distinguishes simple reflex reasoning from what we describe next, is that the output of a circuit evaluation is not evaluated again within the one reasoning response.

There exist other tasks that humans can also do essentially subconsciously, for which it is implausible that simple reflex reasoning is sufficient. Understanding a complex sentence or a picture, or planning how to maneuver out of a room are examples of them. We call them *compound reflex* responses and characterize them as those that are done in several phases of interaction between the NTR and some imagery peripheral. We think of them as having the following general scheme. Given some input the NTR performs a simple reflex task and outputs its deduction to a peripheral. The contents of this peripheral are then added as part of the input for a second simple reflex response, and the result of that is output in turn to the same or some other peripheral. At each phase some immediate consequences of the new input to the NTR are derived and made available so that further consequences can be made. In hearing a sentence, each phrase may conjure up an image which we add to the total picture in our mind's eye. Therefore, as this picture builds up its contents may influence our interpretation of each further phrase. By means of a sequence of such simple reflex responses we attempt to build up an overall picture. In the course of this process we may use additional reflexes to eliminate and correct for inconsistencies.

Compound reflex responses, complex though they may be, we still think of as corresponding to subconscious processes in humans. They are under the automatic control of some algorithms in the peripherals and the NTR, although their course may be influenced by new external inputs or recalled memories. Clearly humans can do reasoning that is even more flexible than this. We call this *Turing reasoning* since, in fact, humans can simulate in their mind any reasoning algorithm consciously, much as universal Turing machines can simulate any particular Turing machine given sufficient work space. In solving nontrivial puzzles, for example, we may go through some systematic reasoning procedure that we have acquired only after extensive study. Although the study of human puzzle solving or conscious reasoning is of intrinsic interest, we believe that among the three modes of reasoning that we have just enumerated it is the furthest removed from the cognitive substrate that we are seeking to uncover and we shall not discuss it in any further detail here.

13.3 Simple Reflex Reasoning

The most basic computation performed by the NTR is a cascade of threshold firings, prompted by some input from the peripherals. Its primary purpose is that of recognition, but, by means of reverse edges or other connections, it can also give rise to a response. Responses that consist of simple circuit evaluations of this nature appear to be limited in their flexibility. They correspond, perhaps, to what are sometimes called "precanned" responses. The number of circuits in an NTR or the brain is clearly limited. For example, it is implausible that humans have a circuit, analogous to an NTR circuit, for evaluating when to answer "yes" for any question stated, say, in English, since it is difficult to imagine how such a circuit could be set up in the brain and updated as new knowledge is acquired. We believe, however, that precanned responses do have a large and important role in intelligence. Reaction time experiments support this view by confirming that humans can perform complex recognition tasks in time that is no more than enough for at most a few dozen neuronal firings.

It is plausible, for example, that associating "56" with the sight of "8 × 7" is performed by humans as a precanned response, at least after extensive training. If the numerals are handwritten, then their recognition, a nontrivial task which presumably is learned in inductive mode, has to precede this, also presumably as a precanned response. Precanned responses may be sufficient for many commonsense tasks that humans can perform without thinking, but which have nevertheless proved difficult to simulate in programmed systems. It is perhaps best to think of a precanned response as a *reflex* in the context of intelligent behavior. A reflex is a program or circuit that under certain input conditions gives rise to a certain output to or through the peripherals. The justification of its validity is the purely empirical one, that it "works" in the world in which it must. How can a system containing a large set of reflexes work without being brittle? Our answer is that each reflex draws its validity of being well tuned to the world from which it has been learned, through the fact that it has been learned in the pac sense. What happens if two reflexes suggest contrary behavior in certain situations? Our solution is that a third reflex that resolves correctly between the two, again in the pac sense, has to be learned or otherwise acquired. Thus at the heart of our theory is the idea that a large number of reflexes learned hierarchically in the pac sense can give rise to a robust intelligent system. The viewpoint that intelligence should be viewed as a large number of largely independent units is held widely and from a great variety of perspectives in artificial intelligence (A. Newell and H.A. Simon, 1972, M. Minsky 1986, A. Newell 1990, R.A. Brooks 1991). Our view is consistent with this general philosophy but contains an additional notion, which is, we believe, central. The additional notion is that it is the very fact that the units are learned rather than programmed, which makes robust systems possible. Thus learning is not the unexplicable obstacle to understanding intelligence, but the key.

We shall discuss in this section four distinct paradigms of reasoning all of which can be viewed as exemplifying simple reflex responses. The first one of these is Boolean circuit evaluation. This can be viewed as an implementation of one of the few systems of logical reasoning known to be computationally tractable, namely Horn clause deduction (H. Levesque 1986). In this, some Boolean implications of the form $x_1 \wedge x_2 \Rightarrow x_3$, $x_1 \wedge x_3 \Rightarrow x_4$ are known, and the truth of some variables is given. The task is to determine which other variables are then implied to be true also. In this case, for example, from the truth of x_1 and x_2 one can deduce the truth of x_3 using the first implication. The second implication then gives the truth of x_4. The restriction on the form of the implications that characterizes Horn clauses and makes them tractable is that no negated variables are allowed to occur. Horn clause deductions, however, are exactly what circuit evaluation performs. If we have a circuit for $x_3 = x_1 \wedge x_2$ and another for $x_4 = x_1 \wedge x_3$, then the prompting of the nodes corresponding to x_1 and x_2 will, within one cascade of firings, also cause the nodes of x_3 and x_4 to fire. Hence, in principle, arbitrary depths of Horn clause deductions can be carried out in a single act of reflex reasoning. The one limitation to this in our system is Principle 3 of the previous chapter that asserts that the number of features of an input that can be presented simultaneously is limited. For this reason some Horn clause deductions may take more than one step and, therefore, take the form of compound reflex reasoning, to be discussed in the next section.

We note that in this, as in all the later paradigms of reasoning, we can invoke Principle 1 of the previous chapter to obtain additional interpretations of the reasoning act. This means that we consider circuit evaluation to be accompanied by reverse flows of information that give rise to images in the peripherals that somehow correspond to the highest level items fired in the NTR. This paradigm therefore corresponds, in human terms, to recalling, being reminded of, making an association, or deducing an implication.

The importance of *context* in human recall and reasoning has been emphasized frequently. A person's behavior in, for example, a restaurant will be governed by detailed previous knowledge relevant to this particular context and will differ from the behavior in, say, a store or office context (M. Minsky 1975, R.C. Schank and R.P. Abelson 1977). We envisage that the role of context in reasoning and recall can be accommodated in the following way. In a restaurant context, the neuroids corresponding to restaurants would be in a high state of activity. Also, the other items that are particularly relevant to restaurants will be those that can be brought most easily nearer to threshold firing when the restaurant neuroids fire. This can be achieved by, for example, having these other items to be conjunctions, with "restaurant" as one conjunct. Hence the restaurant context will effectively bias the items semantically related to it to be the ones that need the least additional stimulus to make them fire.

We now turn to three further paradigms of simple reflex reasoning. These are more complex than circuit evaluation in the sense that they each require a response to a *query*. The queries will be expressed in terms of the quantifiers "there exists" and "for all", denoted by $\exists$ and $\forall$, respectively as in the predicate calculus. These three further paradigms will be distinguished by the subscripts s, m, and w, which denote respectively whether the quantification is over items in the current *scene*, items in *memory*, or items in the *world*. For example,

$$\exists_m \; a \; \text{actor}(a) \; \wedge \; \textbf{US-president}(a) \qquad (13.1)$$

denotes the query whether there is an item in memory corresponding to an individual that has both of the asserted properties. Replacing $\exists_m$ by $\exists_s$ would be the same question, but now restricted to objects depicted in the particular scene currently input. Instead of asking whether you can think of an actor president, you are asked whether there is an actor president in a specific picture. The third category $\exists_w$ refers to the possible existence in the world of an object with the stated attributes. While positive responses to either of the first two queries would normally imply the same for the last, this does not hold for negative responses since the absence of an instance that would confirm some proposition is not taken necessarily as proof of impossibility. On the other hand, based on what we believe to be universal truths, we might deduce that something is indeed impossible in the world. One might, for example, have a belief about a characteristic essential to presidents that is inconsistent with one's beliefs about actors. Although reasoning about universals in this way may be perilous, humans indulge in it freely and some view of this process, even at its least reliable level, needs to be taken. If we see a blue apple, then using some such reasoning we deduce that it is artificial and will desist from taking a bite.

All three kinds of quantification appear to be important. Searching one's memory for examples that satisfy some conjunction of criteria is something humans appear to be able to do remarkably well. It is a fundamental random access task that any plausible model of neural cognition needs to address. Answering similar questions about a particular input, say a picture, is a different task which is also within human capabilities. The category of queries, where quantification is over all possibilities in the world, is also essential. Since earlier we labeled Boolean circuit evaluation as the first paradigm of simple reflex reasoning, we shall label responses to queries using each of these three forms of quantification as the second, third, and fourth paradigms respectively.

Starting with quantifications over memory, which we shall call the second paradigm, consider queries of the form

$$\exists_{\mathbf{m}} \; a_1, \cdots, a_k \; H(a_1, \cdots, a_k).$$

For what values of k and for which classes of expressions H can these be answered by a simple reflex response? By an answer we mean the recollection of objects $a_1, \cdots, a_k$ with the required properties, if these exist in memory, and an answer "no" otherwise.

Two subcases stand out. In the first there is just one object variable and H is a conjunction of unary predicates:

$$\exists_{\mathbf{m}} \; a \; \mathbf{u}_1(a) \wedge \mathbf{u}_2(a) \wedge \cdots \wedge \mathbf{u}_n(a).$$

The task is to identify an item in memory having all n of the properties $\mathbf{u}_1, \cdots, \mathbf{u}_n$. Query (13.1) is an instance of this having $n = 2$. Our observation here is merely that these conjunctions can be accessed by firing the neuroids corresponding to these properties simultaneously, which will cause the nodes representing the conjunctions to fire or come close to firing. If the conjunction used for memorization has more components than that in the query, then this method will bring the desired nodes only closer to the threshold rather than to it. This may be sufficient if the general level of activity is suitably increased so that the threshold will then be exceeded. In practice, we expect that n will equal two or perhaps three. There are various possible algorithmic schemes for achieving the actual retrieval. The important point is that the connections required to channel activity to the sought after conjunction, given activity at the neuroids corresponding to the separate attributes, are already in place in the NTR. Thus the difficult part of the retrieval is solved by the nature of the knowledge representation used.

One specific mechanism for ensuring that a conjunction as in (13.1) can cause the item that it uniquely characterizes to fire, rather than just to come closer to firing, is supplied by the notion of *continuous learning*, which we regard as a crucial aspect of the NTR. By this notion we mean that once neuroids have been allocated for a set of items, these items will learn continuously in inductive mode in terms of each other thereafter. Each input to the NTR causes some item representatives to fire, and others not to. Each such input can be regarded, therefore, as being an input for further refining *every* item that is being learned inductively even if it is a negative example for most of them. Thus the NTR is continuously refining the accuracy of all its concepts as a background activity. Suppose now that this continuous learning happens to do disjunctions by the elimination algorithm. Suppose also that a conjunction for **actor** and **US-president** has been memorized in unsupervised mode. Then after a long period of continuous learning one expects that this conjunction will remain an uneliminated disjunct for just one item in memory. Hence, whenever this conjunction is caused to fire by the query, then so will the sought after item, but nothing else.

The second subcase of this paradigm of retrieval from memory that we would like to mention is one where H is more complex. In particular, assume that it is an L-expression that is closely related to an L-expression already stored. The difficulty here is that, unlike the case of conjunctions, an implemented L-expression may not be accessible by all possible subexpressions of it. For example, if

$$\exists\, a, b, c \;\; \mathbf{r}(a, b) \wedge \mathbf{s}(b, c) \wedge \mathbf{u}(a) \wedge \mathbf{v}(b) \wedge \mathbf{w}(c)$$

is implemented as defined in Figure 11.3 the query

$$\exists_{\mathbf{m}}\, a, b, c \;\; \mathbf{r}(a, b) \wedge \mathbf{s}(b, c)$$

will not retrieve it since the nodes $\mathbf{r}^1 \wedge \mathbf{u}$, etc., will not be made to fire. Queries containing the whole expression will, of course, retrieve it just like a presentation of a scene satisfying it would. Also queries, such as

$$\exists_{\mathbf{m}}\, a, b \;\; \mathbf{r}(a, b) \wedge \mathbf{u}(a) \wedge \mathbf{v}(b),$$

which correspond to a subtree in the representation, would get closer. There appears to be plenty of scope for experimentation to study how humans cope with this issue. Do humans need the labeling predicates that correspond to the u and v that are used in L-expressions, in order to execute such relational queries?

Next we consider the third paradigm of simple reflex responses, those involving queries in which quantification is limited to objects in a particular scene:

$$\exists_{\mathbf{s}}\, a_1, \cdots, a_k \;\; H(a_1, \cdots, a_k).$$

In this case the peripherals will assist in trying all possible ways of identifying $a_1, \cdots, a_k$ with objects in the scene that are viable. If some or all of the a_i are labeled uniquely by predicates (e.g. $\mathbf{u}_i(a_i)$ is part of the conjunction and $\mathbf{u}_i$ does not occur elsewhere in the expression) then fewer ways of identifying them need to be tried. We note that in the $\exists_{\mathbf{s}}$ paradigm a richer set of queries can be processed than in the $\exists_{\mathbf{m}}$ case. They no longer need to be limited to those that relate closely to expressions already memorized. On the other hand, the price that has to be paid is that extra costs may be incurred in enumerating the various permutations of objects in the scene.

A second instance of the paradigm is a query of the form

$$\forall_S \, a \, \mathbf{u}(a) \Rightarrow \mathbf{v}(a).$$

The query asks whether every object in the scene satisfying $\mathbf{u}$ also satisfies $\mathbf{v}$. This can be answered also by examining the objects in turn.

Returning now to the previous paradigm, how can we answer the corresponding query

$$\forall_m \, a \, \mathbf{u}(a) \Rightarrow \mathbf{v}(a) \qquad (13.2)$$

which seems at least equally important? There are at least three distinct mechanisms that may be appropriate. First, if we can access items satisfying $\mathbf{u}$, and there are not too many, then we can retrieve them in turn and test for $\mathbf{v}$. Second, if there is an item $\bar{\mathbf{v}}$ that is the negation of $\mathbf{v}$, then we can test for

$$\exists_m \, a \, \mathbf{u}(a) \wedge \bar{\mathbf{v}}(a)$$

since this is an instance of a conjunction of the form (13.1) discussed earlier. (An alternative to this is the situation in which $\bar{\mathbf{v}}$ is not an item, but there exist items, say $\mathbf{v}_1, \mathbf{v}_2$, and $\mathbf{v}_3$ which exhaust the alternatives to $\mathbf{v}$. For example, if $\mathbf{v} = \mathbf{green}$ then we could have $\mathbf{v}_1 = \mathbf{blue}, \mathbf{v}_2 = \mathbf{red}, \cdots$, etc. and look for $\mathbf{u}(a) \wedge \mathbf{v}_i(a)$ in turn. This is a good strategy, but perhaps beyond what we would typify as a simple reflex.) Finally, if $\mathbf{v}$ stands for an item learned inductively as a disjunction, then the mechanism of continuous learning described above is satisfactory. A positive answer to the current query is equivalent to $\mathbf{u}$ not having been eliminated as a disjunct during the process of learning $\mathbf{v}$.

The fourth and last paradigm that we introduced involves quantification over all possibilities in the world. It relates to statements such as,

$$\forall_W \, a \, \mathbf{grass}(a) \Rightarrow \mathbf{green}(a). \qquad (13.3)$$

We consider that the implementation of such a statement will be a circuit in which whenever the grass nodes fire the green nodes will also fire. Such a circuit can be acquired either by induction or by memorization. In the former case continuous learning of disjunctions suffices yet again, providing yet a third role for this mechanism. In the latter case we can imagine that the statement (13.3) is input as a statement of universal fact, as happens in human instructional contexts, and some mechanism akin to supervised memorization is invoked to create the appropriate circuit.

13.4   Compound Reflex Reasoning

Simple reflex responses are those that, once interactions to achieve attention or to resolve among competing responses are laid aside, consist essentially of one circuit evaluation or a sequence of them applied to different parts of the input in turn. They already capture some complex behaviors. The more remarkable aspects of human commonsense reasoning probably require that these simpler responses be put together into *compound reflex* responses. Our basic view of how this is done is the following. We have peripherals which we call *imagery peripherals*. These may correspond to the same mechanisms that are also variously associated with what is sometimes called working memory. Initially we place information from an input or from the result of an internal computation into these peripherals. Each simple reflex takes the information from them and makes a response that may be considered to be the most immediate deduction that may be made, given the available circuits in NTR and the algorithms that control simple reflex responses. It then adds the result of this response back into the peripherals. As a result of such a succession of simple reflexes, a complex picture will typically evolve in the imagery peripherals.

The imagery peripheral may contain information about several objects and, therefore, the issue of variable binding also arises in this context. L-expressions that label objects distinctly by means of unary predicates appear to be appropriate for this role and therefore issues similar to the ones discussed in Chapter 11 are also relevant here.

The process that guides such a succession of images to a sensible outcome may involve several mechanisms. A philosophical aside is that since simple reflexes are thought of as entirely automatic, there is no reason for any intervention or interruption by the system in the middle of the execution of one such response once initiated. Hence there is no reason for the system to examine or be conscious of the mechanics internal to any one simple reflex. It is, of course, important that the sequence of responses be kept on a productive track. The primary mechanism that might help guide this is one that invokes the most relevant reflexes in the NTR to evaluate each image produced and produce the next. For example, in a car driving situation we may have an image that triggers in the mind a subsequent image suggesting danger. A suitable reflex from the NTR may then be invoked to generate the next image, which has the result of producing evasive physical action. Such a sequence of images may be exactly the appropriate snapshots of points at which new "options" for action are available. At each snapshot information input from the outside or freshly recalled from memory can be incorporated to influence the subsequent snapshots.

This notion of a process being led to a sensible outcome needs further refinement. One aspect of it is consistency checking. When we try to interpret a blurred picture, we may make hypotheses about various parts and use our knowledge and expertise to rule out internally inconsistent interpretations. In addition to learned knowledge we may have innate expertise on, for example, three dimensional space. Our reaction to a glass of water being spilled at our table may be viewed as a compound response built up of expert simple reflexes. Each execution of a simple reflex in such a sequence somehow evaluates the outcome of the previous stage. Clearly there must be some resolving mechanism that picks out one outcome from a range of possibilities. There must exist some control algorithms responsible for this. We suspect that a viable system needs to be in possession of reflexes that are good enough, that in real world situations each execution typically gets pulled toward one outstandingly attractive option, as opposed to being torn between alternatives whose desirability is roughly equal.

This view of reasoning is to be contrasted with the one that emerges from formal logic and has been adapted by some workers in artificial intelligence — namely that reasoning involves some substantial search among alternatives. Indeed, unless one imposes very severe limitations on expressive power, reasoning in systems of formal logic appears to require exponential search and is, to the best of current knowledge, intractable for significant size problems. This, together with various other evidence, has persuaded many other researchers also to seek explanations of human reasoning that are not based on deduction in formal logics. Some of these have much overlap with our view. Craik and Johnson-Laird have developed the notion of "mental models", particularly for the level of reasoning used in puzzle solving. They argue that we reason not in generalities but by imagining particular instances, the models, that satisfy the generalities. Levesque, influenced by the apparent intractability of logical reasoning, has suggested the notion of "vivid" reasoning, where again one argues about single instances having the required properties rather than the truth of the generalities themselves (H. Levesque 1986). In the context of robotics the notion of "reactive" systems has been explored, in which responses to sensed inputs are effected without formal reasoning or representation of state (P. Maes 1990, R.A. Brooks 1991).

Compound reflex responses are under the control of internal algorithms exactly as are the simple reflexes. These algorithms have a tendency to push the computation deterministically forward rather than to allow for all possibilities that have not yet been contradicted. This tendency makes for computational feasibility, but necessarily incurs the cost that the conclusions reached will be colored by the prejudices and irrationalities embodied by the control algorithms. Prejudices and irrationality in general thought processes have been studied experimentally in humans. For example, the irrationality of our commonsense view of probabilities has been amply demonstrated (D. Kahneman and A. Tversky 1973). There is also much evidence that humans do not argue readily from the contrapositive (P.C. Wason 1983). Subjects told that "u implies v" for appropriate predicates instantiated in real world examples, do not easily conclude that "not v implies not u" although this is logically implied. Of course, algorithms for any function in a neural system, including inductive learning, may have similar idiosyncratic tendencies of rushing to judgment.

13.5   Nonmonotonic Phenomena

In the previous sections we argued that several basic processes in commonsense reasoning may be viewed in terms of neuroidal computations. The question arises whether it is possible to understand the essential nature of commonsense reasoning in its entirety from a standpoint as strongly procedural as this. A contrasting alternative approach would be to follow classic logic and to regard the relationship between the reasoning process and the reality about which the reasoning is being performed, as the central and primary concern. In this latter approach reasoning consists typically of operations on formal statements that assert facts of truth, belief, or probability about the world.

We shall argue here that the neuroidal model offers a framework within which the central issues in commonsense reasoning can be studied successfully. The view of reasoning that emerges from this approach has both a strong procedural component by virtue of its emphasis on constructive algorithms, as well as a semantic component that derives from the pac interpretation of learned knowledge. This duality has an important advantage. It recognizes that much knowledge is learned empirically through experience. At the same time it acknowledges that in the course of reasoning, arguments may be invoked that have been acquired by memorization, possibly in the course of instruction from a teacher, and about which no other corroboration is available.

The study of commonsense reasoning has been central to the field of artificial intelligence from the beginning. A number of paradigmatic examples have been used in the literature to illustrate the fundamental difficulties and the various attempts at overcoming them (R. Reiter 1987). We shall use some of these examples below to illustrate the neuroidal approach.

We start by observing that the primitive notion of knowledge in a neuroidal system is that of an *item*, as defined in earlier chapters. The firing of the neuroids that represent an item, such as bird, has a particular meaning that is not identical to any notion in logic or probably in any other system. The firing of the bird neuroids means, at the first level, that "confirmation of birdhood has been established by the current circuits in the system, from the information available to them, which may be incomplete." This characterization is complicated by the fact that there is a duality in the possible meanings of "birdhood" that are allowed here, which reflects the basic duality in the knowledge acquisition process itself. Inductively learned knowledge can be given a precise semantics in terms of the pac interpretation: the bird nodes will fire on those occasions when a true bird is to be recognized. Errors are allowed, but only with small probability on natural examples. Memorized knowledge, on the other hand, typically includes rules, such as "a phoenix is a bird," for which no other supporting evidence is available.

We shall now turn to some particular phenomena of common-sense reasoning that have been identified repeatedly as problematic. Consider the following pair of assertions:

    (i) Quakers are pacifists.

    (ii) Republicans are nonpacifists.

Generalizations of this kind appear to be widely used, even though their meaning may not be clear. Statements of this form may be true by definition, assertions of statistical fact, or nothing more than slogans designed to elicit certain associations in humans. Such rules of thumb appear to be useful, nevertheless, in coping in a complex world. It is reasonable, therefore, that circuits to execute them should be acquirable by memorization, as discussed in §13.3.

A serious problem does arise, however, if invocations of the various assertions in the system lead to contradictory conclusions, as would happen, for example, if the system encountered an individual, such as R. Nixon, who is both a Quaker and a Republican. In classical logic this would be intolerable, but in the neuroidal framework no fundamental difficulty arises. The firing of the nodes corresponding to R. Nixon would cause a chain of firings that would ultimately cause both the pacifist and the nonpacifist nodes to fire, assuming for now that nodes of both kinds, or their equivalents, exist. This information would be output to an imagery peripheral, and the contradiction detected there and then. The contradiction may be resolved by the invocation of appropriate further reflexes, or it may be left unresolved. The detection of contradictions is therefore a natural process within neuroidal reasoning.

Let us now go on and consider the following two assertions:

(iii) Typically high school dropouts are adults.

(iv) Typically adults are employed.

Here the input of a description that satisfies the criterion of high school dropout will, in two stages, lead to the firing of the nodes for employed. This example illustrates that each stage of compound reflex reasoning needs to be done on the totality of the information available, rather than merely on the output of the stage before. As discussed in the previous section, we envisage that an imagery peripheral stores both the original input as well as the deductions made at intermediate stages of the reasoning. Hence when the employed nodes fire, the reference to high school dropout is still available, and any implied contradiction can be detected.

The two examples just presented are of limited interest, perhaps exactly because simple procedural solutions appear to suffice for them. They do serve, however, to introduce a more general and fundamental class of reasoning processes, that are variously described as *nonmonotonic* or *default* reasoning. These are characterized by situations in which reasoning is performed and conclusions arrived at when some relevant information is still unavailable. For example, if we know that Nixon is a Quaker, then we may draw conclusions about his pacifism. This is the default assumption to be made about Quakers in accordance with the first of the two rules. The reasoning becomes nonmonotonic when, as a result of a new piece of information, such as that Nixon is a Republican, a previously reached conclusion is withdrawn or changed.

We have emphasized that neuroidal systems need to tolerate having pieces of knowledge that are inconsistent with each other. An important feature of nonmonotonic phenomena, however, is that they occur even in contexts in which one can argue that no inconsistencies exist at all. The most celebrated example of nonmonotonic reasoning is specified by the following pair of statements, that are to be considered in sequence:

(v) Tweety is a bird.

(vi) Tweety is a penguin.

On seeing the first we draw the conclusion that "Tweety flies," if we have available the default rule that "birds fly." Once we see the second statement, we not only withdraw the conclusion just reached, but arrive at one that contradicts it. Therefore, we have a phenomenon in which the addition of a premise that superficially does not appear to contradict the previous premises, leads to a conclusion that directly contradicts the previous conclusion. This phenomenon is inconsistent with classical logic on the surface. Numerous approaches have been explored to find a theory that accommodates it (J. McCarthy 1980, M.L. Ginsberg 1989) and many obstacles encountered.

One approach is based on the observation that default reasoning usually treats an unconfirmed proposition as a false proposition. The proposition that "Tweety is a penguin," for example, is unconfirmed and therefore assumed to be false prior to the availability of statement (vi). This approach has been characterized by the so-called Closed World Assumption, that can be paraphrased as asserting that "what I don't know to be true I can assume to be false."

We shall argue here that the neuroidal approach provides a satisfactory view of nonmonotonic reasoning almost immediately, without needing any such additional assumptions. In particular, the combination of a positive representation of items together with a pac interpretation of their semantics provides an intuitively plausible explanation of how reasoning with incomplete information can be made intellectually defensible.

As we suggested earlier, an item such as penguin corresponds roughly to the predicate "confirmed to be a penguin by the existing circuits." The nonfiring of these penguin nodes corresponds to "not confirmed to be a penguin by the existing circuits." (There may be represented, of course, a separate item "confirmed to be a nonpenguin by existing circuits," but we shall assume for now the simplest case in which items do not typically have their negations represented explicitly like this.)

The crucial point is that the main dichotomy in the neuroidal circuit is between "confirmed to be a penguin" and "not confirmed to be a penguin," rather than the dichotomy "is a penguin" versus "is not a penguin," that characterizes the world. The main difference is that the former accommodates incomplete information seamlessly, while the latter does not. The consequence is that learning in the circuit takes place in terms of variables that take on values "confirmed" or "not confirmed", and therefore allow for incomplete information. The realm of incomplete information *is* the natural domain in which learning takes place. If a recognition circuit for flies is learned inductively, the circuit will process the values of such variables as "confirmed to be a bird" and "confirmed to be a penguin." If the resulting circuit makes **flies** fire whenever **bird** fires, and **penguin** does not, then it must be the case that for natural inputs that make **bird** fire but fail to confirm penguinhood, it is appropriate in the pac sense that **flies** fires also. The question of whether such a circuit is appropriate is more than a simple statistical question about the world, since it depends both on the functions of the existing circuits in the system, as well as on the probability distribution of the examples experienced by the system. The particular circuit in question may be a valid one for an urban inhabitant, who may find that the circuit shown in Figure 13.1(a) "works." For an Antarctic explorer this circuit may not produce the right result for many natural examples. The circuit shown in Figure 13.1(b) with nodes for the item "confirmed not to be a penguin," may be more appropriate in that environment. The point is that, in either case, learning takes place from examples in which the truth of "is a penguin" may or may not be determined. The circuit learned will need to be valid according to the pac criterion on new examples in which also this predicate may or may not be determined. To illustrate this further we note that if the inputs are all visual, it may be that the circuit in Figure 13.1(a) *is* quite adequate for the Antarctic explorer after all. If penguins are always visually identifiable as penguins, then birds of unidentifiable species can be assumed not to be penguins.

To summarize, instead of needing a hypothesis, such as the Closed World Assumption, we now have a derived principle that explains why learned information can be applied with some confidence to situations with incomplete information. This may be called the Closed Mind Principle. Rather than saying "what I

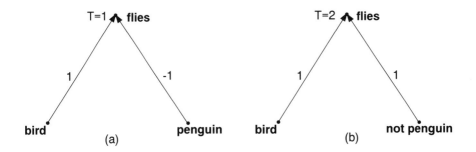

Figure 13.1. Two circuits that embody alternative treatments when information about penguinhood is incomplete.

don't know I can assume to be false," it says "what I don't know I am fairly sure I can do without knowing."

Chapter 14
More Detailed Neural Models

14.1 Implementing Vicinal Algorithms

In previous sections we described a number of algorithms for solving our various idealizations of simple cognitive tasks. The algorithms were all vicinal in the sense that whenever two nodes wished to communicate in some manner either they had a common neighbor and the necessary communication could be realized through that node, or there was none, but this mattered little since the replication of nodes having the same intended function ensured that a sufficient fraction of the parallel attempts at communication succeeded.

Vicinal algorithms have the clear conceptual advantage that they are simple to describe and understand. The difficulties that may arise in realizing communication through an arbitrary network have been factored out. This is achieved, however, not without cost. Some apparently severe constraints appear to be needed if the algorithms described in the previous chapters are to be implemented exactly as described. First, they appear to require that any node be reachable from any other with high probability via just one intermediate node. Second, the algorithms as described all assume that the weight of an edge can be as high as the neuroid's threshold. A signal in just one incoming edge is then sufficient to make the neuroid fire.

The purpose of this chapter is to show that all the vicinal algorithms described can be implemented, though less directly, on much wider classes of graphs and with neuroids having smaller weights relative to their threshold. We interpret this as providing further support to our central thesis that the simple functions implemented in the previous chapters are within the computational capabilities of biological neural systems. We note that the models described in §14.2 and §14.3 are somewhat detailed. These are the ones that attempt to go the furthest toward modeling the cortex itself. The last two models, described in §14.4 and §14.5 are less realistic, and we include them here only to make some theoretical points about the algorithmic possibilities. We emphasize, however, that even in the case of §14.2 and §14.3, while the parameters we assume are not inconsistent with current knowledge, there is no experimental evidence to date to suggest that the mechanisms or representations we describe there are, indeed, used in the brain.

The implementations in §14.2 and §14.3 may be viewed as relating two hypotheses about the cortex to each other and to the implementability of vicinal algorithms. The primary hypothesis is the following: At least a certain fraction, say in the range 0.1% to 1.0%, of those neuron pairs that synapse on each other have the further property that it is possible for one of the pair to be caused to fire at a significantly different rate from the background as a result of the other firing at such a different rate. This hypothesis will be shown to be sufficient for supporting vicinal algorithms. The implementations described imply a secondary hypothesis that provides a detailed theory of how the primary hypothesis might be realized. This secondary hypothesis suggests that the neuron pairs that have the potential for such an exceptionally large influence, are those few that make a significantly above average number of synapses on each other. It is quite sufficient for our purposes, of course, that the primary hypothesis be true, without the secondary one also being true.

Randomness assumptions similar to those used in §14.2 and §14.3 have been used previously to compute the probability of synapsing between pairs of neurons in close proximity (V. Braitenberg 1978, M. Abeles 1991). Here we shall use them to analyze the related phenomenon of multiple synapsing, which is at the heart of the relationship we demonstrate between the two hypotheses.

The vicinal algorithms described in previous chapters are all based on two stereotypes of communication, JOIN and LINK, which implement storage allocation and associations, respectively. We shall show that both of them can be simulated, in a certain probabilistic sense, on the various new models we consider in this chapter. It will then follow that all our vicinal algorithms can be implemented on each of these classes of graphs.

The operation JOIN(x, y) has the two items x and y as arguments and performs the following memory allocation task. For items x and y for which storage has been previously allocated to node sets $\tilde{x}$ and $\tilde{y}$, respectively, it allocates a set $\tilde{z}$ of about r nodes and changes some weights so that whenever $\tilde{x}$ and $\tilde{y}$ fire in the future, so will $\tilde{z}$ (but if just one or if neither of the sets $\tilde{x}$ or $\tilde{y}$ fires, then $\tilde{z}$ will not be made to fire, at least not as a result of these weight changes alone).

The operation LINK(x, z) is also performed on two items, x and z in this case, that have storage already allocated. It can be considered as one establishing an association between the two items. It will cause weight changes in such a way that whenever $\tilde{x}$ fires in the future so will $\tilde{z}$ (and if $\tilde{x}$ does not fire then $\tilde{z}$ will not either, at least not as a result of these weight changes alone).

We note that these statements about the sets $\tilde{x}, \tilde{y}, \tilde{z}$ firing should not be interpreted as implying that at any time either every member or no member of these sets fires. If close to all of $\tilde{x}$ fire then $\tilde{x}$ will be considered to fire and the item x will be considered to have been recognized. On the other hand, if only a small fraction, say less than νr of the nodes fire, for ν substantially less than one, then $\tilde{x}$ will be considered as not firing and x as not being recognized. Note also that in order to simplify the discussion, we shall throughout this chapter refer rather loosely to the expected number of nodes that have a certain property, as if it were the actual number. We shall also make without comment various other approximations that hold for the range of parameters of interest but not necessarily in general.

As described, all the circuits that are established are purely excitatory. They can be adapted, however, to have inhibitory effects on the target nodes as required if there is a sufficient density of neurons having inhibitory effects that they can be called upon to form an additional penultimate link in the chains of neurons. Inhibitory neurons in the cortex are believed to have only short range connections. As long as they can be found near every pyramidal neuron, they can be used to invert the effect on that neuron of any long range pyramidal axon passing in its vicinity.

14.2   A Laminar Model

First we shall discuss a graph model that is based on the availability both of dense local connections as well as of sparser global connections. A special characteristic of local connections in the cortex is that any two neurons physically close together, say within a fraction of a millimeter, have a significant probability of synapsing on each other. We shall define a parameter ρ to be the expected number of synapses that a neuron makes on another that is within a certain distance of it. A high value of ρ is clearly favorable for the network to have the potential to disseminate information from one point in many directions. This favorable density, however, exists in the cortex only locally, and is not enough by itself to support random access tasks. Long distance connections enjoy what can be viewed as the reverse combination of attributes. While they support random access tasks by connecting distant parts of the cortex directly, by themselves they form only a sparse network. We will, therefore, need to combine the best of both in order to achieve our aims.

The major goal that we want to achieve with this model, is to implement vicinal algorithms with neuroids having smaller weights than hitherto assumed. We define the parameter α to be the minimal number of weights that can add up to the threshold needed for a neuroid to fire. The algorithms described in previous sections assume $\alpha = 1$ implicitly, since weight $w_{ji} = 1$ and threshold $T_i = 1$ are allowed simultaneously. Here we shall show that much less stringent requirements are sufficient, such as $\alpha = 5$, with apparently realistic parameters. The extreme possibility, that $\alpha = 1$

characterizes major computations in the cortex, is not supported by current experimental evidence, but has not been completely excluded either. Some of the vicinal algorithms described in previous chapters for $\alpha = 1$ require only a fraction of one percent of the synapses to have such high values. It is not surprising, perhaps, that experiments to date have not detected any such synapses even if they exist. Synapses corresponding to $\alpha = 5$ have, however, been found in the cortex, and it is, therefore, significant that our algorithms can be implemented on a model with α in this range. We note also that it is conceivable that the physical reality corresponds to $\alpha = 5$, say, but the computations performed have the character of a smaller α. This would be the case, for example, if the background random firings in the cortex maintained neurons close to the threshold voltage, so that it is enough for just one or two synapses from purposeful neighbors to push any one neuron over the edge.

The simplest rendering of the model that we use here has two further attributes that are reminiscent of the cortex. First, it is *laminar*, in the sense that the neurons are most conveniently regarded as being organized in layers. Second, it allows for cortex to be partitioned into *areas*, where each pair of areas may be connected in one or both directions. In §14.3 and §14.5 we also consider the *columnar* aspect of the cortex.

The neuroids of each area A are partitioned into two layers $\bar{A}$ and A^*. The former *basis layer* contains the neuroids that represent the items most directly. They have local connections to the *secondary* layer A^*. A^* has long range connections shared in equal numbers among θ distant areas. These distal axonal branchings synapse on the neuroids of the basis layers of these distant target areas. We shall assume that there is a parameter χ that equals both the number of synapses on the local connections, as well as the number on the distal connections, and that its value is 20,000 as given in §2.3. We shall also use the value that is given in that section as the estimate of η, the number of neurons in each cubic millimeter of cortex. This value also happens to equal 20,000. The analysis treats the local and distal connections somewhat differently, and we shall consider these two cases in turn.

The fact that we wish to establish before analyzing the implementations of JOIN and LINK in detail, is that the primary hypothesis stated in §14.1 follows from the secondary one under some reasonable assumptions. We define β to be the expected number of nodes in A^* that synapse with any one node in $\bar{A}$ at least α times. We view β as an amplification factor since for any small random number of the nodes in $\bar{A}$ that fire, about β times as many may be caused to fire in A^*. Showing that an amplification factor of β = 100 is possible, for example, speaks to the primary hypothesis by asserting that a node can by itself cause to fire one hundred of its neighbors, which is a fraction of one percent of their total number.

In the analysis of the local connectivity we shall make the following simplifying assumptions. Suppose that each local axonal branching occupies an effective volume of σ cubic millimeters. One possible estimate is that $\sigma = 0.5$, corresponding approximately to a sphere of radius 0.5 mm, which would have volume $(4/3)\pi(0.5)^3 \approx 0.52$ cubic millimeters. Using the estimate that there are $\eta = 20{,}000$ neurons in each cubic millimeter of cortex, we also suppose that this space of σ cubic millimeters is shared by the dendritic trees of $\psi = \eta\sigma$ neurons in A^*. Finally, we also suppose that each such local axonal branching forms $\chi = 20{,}000$ synapses with neurons in A^*. If the synaptic connections are random, then any fixed neuron in $\bar{A}$ will form its χ local axonal synapses randomly with the $\psi = \eta\sigma$ neurons in A^* sharing the same volume. Hence the expected number of synapses with any one such neuron is $\rho = \chi/\psi = \chi/(\eta\sigma)$ which, for the numerical parameters assumed, equals $20{,}000/(20{,}000\sigma) = 1/\sigma$. These local connections are illustrated schematically in Figure 14.1 below.

We now wish to estimate $p(k, \rho)$, the probability that a fixed node in $\bar{A}$ synapses with a fixed node in A^* exactly k times. This is given by the Poisson distribution

$$p(k, \rho) = e^{-\rho}\frac{\rho^k}{k!} \tag{14.1}$$

where $e = 2.71828...$ is the exponential constant. (In the terminology of gambling, this expression approximates the probability of winning exactly k times when playing a series of m independent games in which the expected total number of wins is ρ, and m is very large. In this interpretation each game corresponds to one

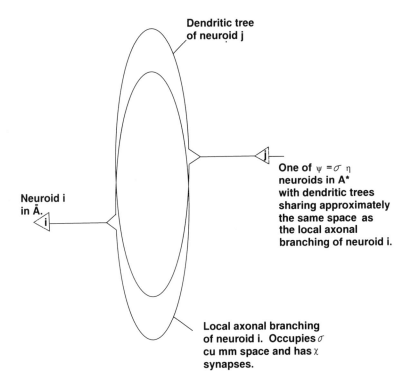

Dendritic tree of neuroid j

One of $\psi = \sigma \eta$ **neuroids in A* with dendritic trees sharing approximately the same space as the local axonal branching of neuroid i.**

Neuroid i in Ā.

Local axonal branching of neuroid i. Occupies σ **cu mm space and has** χ **synapses.**

Figure 14.1. Schematic diagram of connections in the laminar model, between the local axonal branching of a neuroid i in the base layer $\bar{A}$ and the dendritic tree of a neuroid j in the secondary layer A^*. The same diagram has an alternative interpretation in terms of distal connections. In that case i is interpreted as being in the secondary layer A_1^* of one area, the axonal branching illustrated is its distal one, and j is in the base layer $\bar{A}_2$ of the distant area.

of the synapses of the $\bar{A}$ neuron, and a win corresponds to this synapse being made on the fixed A^* neuron rather than some other neuron.) The quantity we really want is β, which we define as the expected number of nodes in A^* that synapse with a fixed node in $\bar{A}$ *at least* α times. To obtain this we shall sum $p(k, \rho)$ for all $k \geq \alpha$, and multiply this by ψ, which is the number of candidate nodes in A^*. Hence, putting $\psi = \chi/\rho$ we get

$$\beta = \frac{\chi}{\rho} \sum_{k=\alpha}^{\infty} e^{-\rho} \frac{\rho^k}{k!}. \tag{14.2}$$

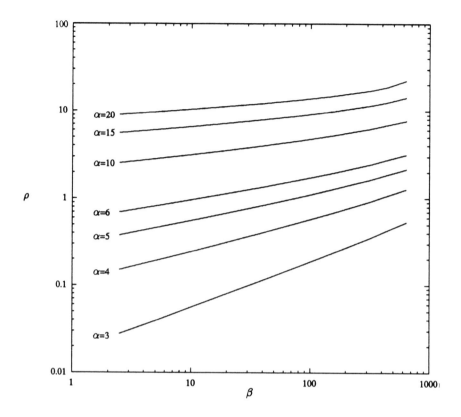

Figure 14.2. Each curve illustrates one choice of α, the ratio of threshold to maximal edge weight. The horizontal axis β measures the amplification factor, which is the expected number of neuroids with which a single neuroid synapses at least α times in the model described. The vertical axis measures ρ, the expected number of synapses between two neuroids. Both axes are on a logarithmic scale.

Figure 14.2 above shows solutions to this equation with $\chi = 20,000$ for $\alpha = 3, 4, 5, 6, 10, 15$ and 20. It shows, for example, that for the estimate $\sigma = 0.5$ and hence $\rho = 1/\sigma = 2$, amplification by β in excess of 100 can be achieved by realistic values of α such as 6. With more modest estimates of ρ one can also achieve significant magnification. For example, with $\alpha = 4$ the value $\rho = 0.5$ would give $\beta > 70$, and the value $\rho = 0.25$ would give $\beta > 10$.

The important point is that even when α is significantly greater than the average number of synapses ρ, each neuron in $\bar{A}$ can have its influence in A^* amplified by a large factor β. The amplification is afforded by the fact that the random connections ensure that the influence of each neuron on a small lucky subset of its neighbors is much larger than its influence on its typical neighbor. The benefit of random connections that is exploited here is that synapsing at well above the average rate is ensured for a small but sufficient fraction of the neurons in A^*.

In obtaining these figures several numerical estimates of the parameters of the cortex were made, as were some further assumptions. The conclusion was that for some parameters ρ, ψ, the expression

$$f(\alpha) = \psi \sum_{k=\alpha}^{\infty} e^{-\rho} \frac{\rho^k}{k!}$$

gives the number of neurons on which one other neuron synapses at least α times. The general form of this expression is plausible if one assumes that axonal branchings and dendritic trees that fill approximately the same volume, synapse randomly. Our analysis is consistent with a reasonably large range of values of ρ and α and it remains to be confirmed that the corresponding parameters of cortex are indeed within these ranges. In §14.3 we will extend the analysis to an even broader range, by incorporating some further plausible assumptions about the cortex.

We go on now to analyze how this amplification phenomenon can be exploited when implementing JOIN. We suppose that two items x and y are represented by sets of r nodes each, $\tilde{x}$ and $\tilde{y}$, respectively. We wish to allocate storage to represent $z = x \wedge y$. We shall assume that $\tilde{x}$ and $\tilde{y}$ are both in area A_1, and z is to be allocated in an area A_2 to which there are connections from A_1. This arrangement will also allocate $x \wedge y$ in each of the other areas to which A_1 projects. We could equally have $\tilde{x}, \tilde{y}$ in distinct areas, both projecting to A_2, in which case an allocation will be only in areas to which the areas of both $\tilde{x}$ and $\tilde{y}$ project.

The nodes $\tilde{x}$ and $\tilde{y}$ lie in the base layer $\bar{A}_1$ of A_1. We define $\tilde{x}^*, \tilde{y}^*$ to be the set of nodes in layer A_1^* to which there are at least α synapses from $\tilde{x}, \tilde{y}$, respectively. The algorithm JOIN will then

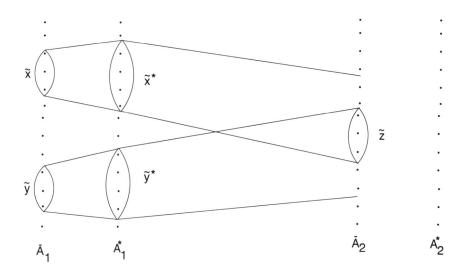

Figure 14.3. Schematic illustration of the implementation of JOIN in the laminar model. If $\tilde{x}$, $\tilde{y}$ represent x, y in layer $\bar{A}_1$ of area A_1, then $\tilde{z}$ will represent $z = x \wedge y$ in layer $\bar{A}_2$ of area A_2.

allocate $\tilde{z}$ to be exactly those nodes in the base layer $\bar{A}_2$ of the second region A_2, that are synapsed at least α times from *both* of $\tilde{x}^*$ and $\tilde{y}^*$. We note that by modifying the algorithm suitably we could reduce this requirement from α to $\alpha/2$ and thus allow for an even wider range of consistent parameters.

Assume that the number of synapses on the distal axonal branching of a neuroid is the same $\chi = 2 \times 10^4$ as on the local branchings, and that each dendritic tree that receives these distal connections occupies the same σ cubic millimeters of volume, as one that receives local connections. Then Figure 14.1 captures this case also. The analysis used to derive equation (14.2) will, therefore, apply in relating the number of synapses of an A_1^* neuroid on an $\bar{A}_2$ neuroid, just as it relates the number of synapses of an $\bar{A}_1$ neuroid on an A_1^* neuroid. For simplicity we shall assume that A_1^* and $\bar{A}_2$ each contain N nodes. Given any ρ and α, a value of β is implied by Figure 14.2. Now if the r nodes of $\tilde{x}$ fire in $\bar{A}_1$, then about $r\beta$ nodes will fire in A_1^*, provided $r\beta$ is substantially smaller than N. Among these only a fraction θ^{-1} will have axons going to $\bar{A}_2$, and the remainder go to other regions. Hence the number of nodes in $\bar{A}_2$ that will be caused to fire by $\tilde{x}$ will have expectation about $(r\beta^2)/\theta$, if this number is sufficiently smaller than N.

If we wish that $\tilde{z}$ be an equal citizen with $\tilde{x}$ and $\tilde{y}$ and, therefore, have expected size r, then we need that the probability that a randomly chosen node in $\bar{A}_2$ be caused to fire by both $\tilde{x}$ and $\tilde{y}$ firing alone, be equal to r/N if the number of nodes in $\bar{A}_2$ is N. Since the probability that a random one of the N nodes in $\bar{A}_2$ is made to fire by $\tilde{x}$ is $(r\beta^2)/(N\theta)$, and the same expression holds for $\tilde{y}$, it follows that we need that

$$\left(\frac{r\beta^2}{N\theta}\right)^2 = \frac{r}{N}. \tag{14.3}$$

Putting $\kappa = N/r$, which roughly equals the number of items that can be stored in area A_2, gives

$$\kappa\theta^2 = \beta^4. \tag{14.4}$$

We claim that in the full range of values of κ and θ that we envision, the required value of β can be achieved for reasonable values of the parameters ρ, α and ψ on which β depends. What ranges of values of κ and θ are reasonable? If the total number of neurons in cortex is 10^{10} and the replication factor r is of the order of 10^2, then we expect to represent at most 10^8 items. Also, if there are ϕ areas of equal size, then $\phi\kappa = 10^8$. To reconcile this with $\kappa\theta^2 = \beta^4$ we need $\phi\beta^4 = 10^8\theta^2$ or $\beta = 100(\theta^2/\phi)^{1/4}$. Now, if we consider, for example, the case in which each area is connected to the square root of the total number of areas, then $\theta = \phi^{1/2}$ and we get $\beta = 100$. How large a multiplicative factor can $(\theta^2/\phi)^{1/4}$ become in other cases? It seems that it cannot exceed unity by much. For example, if $\phi = 1000$, i.e. we have 10^3 areas of 10^7 neurons each, and if each is connected to $\theta = 100$ others, then this factor becomes $10^{1/4} \approx 1.78$. At the opposite end of the range it is conceivable to have $\phi = 1000$ and $\theta = 2$, in which case $\beta = 100/250^{1/4} \approx 25$ is sufficient.

If equations (14.2) and (14.3) are satisfied then Algorithm 7.2 can be supported on this neuroidal scheme. The weights of the neuroids in A_1^* are never modified, essentially being fixed to a fraction $1/\alpha$ of the magnitude of the threshold T. It is the neuroids in $\bar{A}_2$ that are modified and execute the algorithm. The effect of A_1^* is simply to create much enlarged sets $\tilde{x}^*$ and $\tilde{y}^*$ to represent $\tilde{x}$ and $\tilde{y}$ in order to increase the number of synapses that are made in $\bar{A}_2$.

In what sense have we implemented an algorithm for unsupervised memorization? Clearly the circuit constructed will be such that whenever $\tilde{x}$ and $\tilde{y}$ fire, so will $\tilde{z}$. Is there a possibility that the weight changes made in $\tilde{z}$ will cause it to fire on future occasions even when not both of $\tilde{x}$ and $\tilde{y}$ are firing? A worst case for this is that of having just one of $\tilde{x}$ or $\tilde{y}$ firing simultaneously with a number of spurious sets $\tilde{x}_1, \cdots, \tilde{x}_k$. We will show that, if k is small, then it is unlikely that too many members of $\tilde{z}$ will be caused to fire in this way. If we regard the item z as Boolean, then we can consider that its truth is indicated only when a certain fraction νr of its nodes fire, for some constant $0 < \nu < 1$. We can then deduce that spurious inputs will not make z appear to be true when it is not, except very rarely.

Suppose, therefore, that after the circuit for $z = x \wedge y$ has been established by means of Algorithm 7.2, $\tilde{x}, \tilde{x}_1, \cdots, \tilde{x}_k$ in $\bar{A}_1$ all fire, but $\tilde{y}$ does not. These will cause the $r\beta$ nodes $\tilde{x}^*$ in A_1^* to fire, as well as a further $kr\beta$ nodes in A_1^*. Now among these latter $kr\beta$ nodes, $kr\beta/\theta$ will have axons going to $\bar{A}_2$, and each of these will cause β nodes to fire in $\bar{A}_2$. The probability that these $r\beta^2 k/\theta$ nodes include exactly i that are in $\tilde{z}$ is no more than

$$\binom{r\beta^2 k/\theta}{i} \left(\frac{r}{N}\right)^i \leq \left(\frac{r^2\beta^2 k}{\theta N}\right)^i \frac{1}{i!} \qquad (14.5)$$

$$\leq \left(\frac{rk}{\sqrt{\kappa}}\right)^i \frac{1}{i!}$$

after substituting $\kappa = N/r$ and $\kappa = \beta^4/\theta^2$ from $(14.4)^{32}$. If we assume that $N \geq 10^7$, then $rk/\sqrt{\kappa} \leq 1$ provided $rk \leq 10^{7/2}/r^{1/2}$. For this $r = 50$ and $k = 8$ is sufficient. Hence the bound in (14.5) is then at most $1/i!$, which in turn is $(10!)^{-1} \leq 10^{-6}$ for $i \geq 10$. Hence if $r = 50$, for example, then a random set of $k = 8$ items $x_1, \cdots, x_8$ firing will cause no more than 10 members of $\tilde{z}$ to fire,

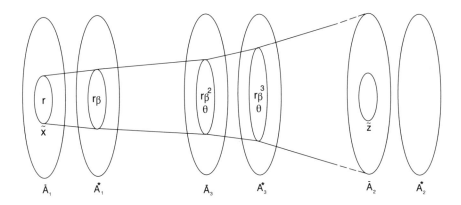

Figure 14.4. Schematic illustration of an implementation of LINK from $\tilde{x}$ in area A_1 to $\tilde{z}$ in area A_2. Each large ellipse represents one layer of one area.

except with probability 10^{-6}. We conclude that, under these conditions, the possible unwanted side effects of this implementation of JOIN will not be harmful.

Now we turn to how LINK is to be implemented. We have items x and z in areas A_1 and A_2, respectively. We need to set up a circuit such that at any later time whenever $\tilde{x}$ fires so will $\tilde{z}$.

Several implementations are possible. We will consider one here in which there is a third area A_3 such that A_1 projects to A_3 and A_3 projects to A_2. Each area is composed of two layers, the base layers $\bar{A}_1, \bar{A}_2$ and $\bar{A}_3$ and the secondary layers A_1^*, A_2^*, A_3^*. We assume that all the parameters are exactly as in the implementation of JOIN except now the neurons of $\bar{A}_2, \bar{A}_2^*$, as well as A_1^* all act as relay nodes, just as those in A_1^* did in JOIN. Their weights are fixed to a fraction $1/\alpha$ of their firing threshold and are never modified.

In this configuration the r neurons of $\tilde{x}$ in $\bar{A}_1$, can cause to fire an expected number of about $r\beta$ nodes in A_1^*, $r\beta^2/\theta$ nodes in $\bar{A}_3$, and $r\beta^3/\theta$ nodes in A_3^*, provided each of these forms a small fraction of the respective areas. Further, each of the latter will synapse at least α times with each of β/θ nodes in $\bar{A}_2$. (Note that the less stringent requirement that α synapses altogether be formed on sufficiently many nodes in $\bar{A}_2$ would be enough here.) We need that with high probability each of the nodes in $\tilde{z}$ be amongst those that are synapsed at least α times by at least one of the $r\beta^3/\theta$ nodes in A_3^*, for then Algorithm 8.1 or 8.2 can be executed for the $\tilde{z}$ nodes. Now the probability that all $r\beta^4/\theta^2$ such opportunities miss any one fixed $\tilde{z}$ node is

$$\left(1-\frac{1}{N}\right)^{r\beta^4/\theta^2} = \left(1-\frac{1}{N}\right)^{\frac{Nr\beta^4}{N\theta^2}} \le e^{-\frac{r\beta^4}{N\theta^2}}. \qquad (14.6)$$

If we want this to be k^{-1} for some $k > 0$, then we need $(\log_e k)\kappa\theta^2 = \beta^4$, which is the same as equation (14.4) with the exception of the $\log_e k$ factor. If we want 98% of the nodes of $\tilde{z}$ to fire then $k = 50$ and $\log_e k \approx 3.91$. We can accommodate equations (14.4) and (14.6) in the same system in many ways. In particular, since the parameters θ and β associated with the special area A_3 may be different from those of other areas, we could distinguish them as different parameters having slightly different values and hence have the corresponding modifications of (14.4) and (14.6) satisfied simultaneously. For example, the θ in (14.6) could be replaced by $\theta_1 \approx \theta/2$ to achieve this result for $\log_e k \approx 4$.

We note that in the representation described here, in the base layer $\bar{A}$ of an area A we use grandmother cells, while in the secondary layer A^* we have essentially population coding. Furthermore, we could change our viewpoint and regard the A^* nodes as being the true representatives of the item. The function of the grandmother cell coding in $\bar{A}$ is to help assign storage and to make the firing conditions for the representations of the different items distinct. The function of population coding in A^* is then to allow the toleration of higher values of α than would be otherwise possible.

14.3   A Columnar Model

The analysis given in §14.2 presupposed that the r neuroids representing an item were uniformly scattered at random in the appropriate layer or area. This assumption is clearly attractive from the viewpoint of fault-tolerance since it minimizes the negative effects of neuroid damage within any one locality.

Various arguments suggest that one should also consider representations where items are stored with more locality. One source of motivation is the observation made by many neuroanatomists that the cortex is not laterally uniform, but divided up into vertical units, that are sometimes called *columns*. Estimates of the size of these units vary from the order of tens of neurons to tens of thousands. Also there is no concensus on how much more dense the connections internal to one such column are as compared with the connections between distinct columns. As we shall see, a columnar organization with some locality properties has some advantages for realizing random access tasks.

We shall describe a very simple way of modeling locality. Analysis will show that locality does make the model more robust than the simple laminar model, in allowing even higher values of α to suffice in supporting vicinal algorithms. For example, a value of $\alpha = 15$ now suffices where $\alpha < 5$ was needed in the previous section.

We need to define locality as it affects both the representation of an item in the base layer $\bar{A}_1$ as well as in the long range connections from A_1^* to the base layer, say $\bar{A}_2$, of another area. Our model uses a parameter ξ that expresses the degree of locality, and takes on some positive value, such as $\xi = 5$. We identify a *column* here as consisting of roughly $\psi = \eta\sigma$ neurons in each of the two layers, corresponding to the volume occupied by the local or distal axonal branchings of a neuron. To capture the columnar model at the local level we assume that an item is represented not by r random neurons in $\bar{A}_1$, but by r/ξ groups of ξ neurons each, where each such group is within one column but the r/ξ groups are in randomly chosen columns of $\bar{A}_1$. At the global level we assume that the $\eta\sigma$ long range axons from any column of A_1^* go to only β_ξ distinct columns, shared equally among θ areas, of which $\bar{A}_2$ is one. This choice is made here mainly for ease of analysis, and for a value of β_ξ to be defined below (14.8). In a more detailed analysis we would give the parameter ξ different names and values in the local and gobal contexts, but, for simplicity, we do not do that here.

We shall adapt our analysis of JOIN to this model. First we

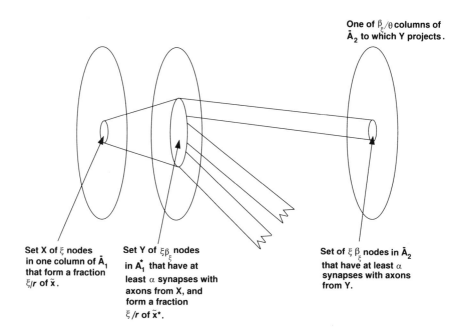

One of β_ξ/θ **columns of** $\bar{A}_2$ **to which Y projects.**

Set X of ξ **nodes in one column of** $\bar{A}_1$ **that form a fraction** ξ/r **of** $\tilde{x}$.

Set Y of $\xi\beta_\xi$ **nodes in** A_1^* **that have at least** α **synapses with axons from X, and form a fraction** ξ/r **of** $\tilde{x}^*$.

Set of $\xi\beta_\xi$ **nodes in** $\bar{A}_2$ **that have at least** α **synapses with axons from Y.**

Figure 14.5. Schematic illustration of a part of a circuit that implements JOIN in the columnar model. Each large ellipse represents a layer in one column.

consider the synapsing of the $\bar{A}_1$ nodes on the A_1^* nodes. The probability that any set of ξ nodes in a column in $\bar{A}_1$ synapse with a fixed node in the same column in A_1^* exactly k times between them is

$$\rho(k, \xi\rho) = e^{-\xi\rho}\frac{(\xi\rho)^k}{k!}, \qquad (14.7)$$

by analogy with (14.1). This follows since the expected number of synapses between the ξ chosen nodes in $\bar{A}_1$ and the one chosen node in A_1^*, is now ξ times greater than in the case expressed by (14.1), in which just one node in $\bar{A}_1$ was considered. Let us define β_ξ to be r^{-1} times the expected number of the nodes $\tilde{x}^*$ in A_1^* that are synapsed at least α times by the representatives of an item $\tilde{x}$ in $\bar{A}_1$. Since there are $\psi = \eta\sigma = \chi/\rho$ nodes in A_1^* that are candidate targets for each of the r/ξ groups of ξ nodes in $\bar{A}_1$, the expected number of nodes in $\tilde{x}^*$ is

$$r\beta_\xi = \frac{r}{\xi} \times \frac{\chi}{\rho} \sum_{k=\alpha}^{\infty} e^{-\xi\rho} \frac{(\xi\rho)^k}{k!}.$$

or

$$\beta_\xi = \frac{\chi}{(\xi\rho)} \sum_{k=\alpha}^{\infty} e^{-\xi\rho} \frac{(\xi\rho)^k}{k!}. \tag{14.8}$$

Thus the number neuroids that can be caused to fire in A_1^* in this columnar model is given by the same expression (14.2) as before, except that ρ is replaced by $\xi\rho$. In other words, when the same value of β is to be realized in both models, a higher value of α can be tolerated in the columnar model, and this value can be read off from Figure 14.2. For example, $\beta_\xi = 100$ can be achieved with $\alpha = 15$ if $\xi\rho \approx 9.1$, for which $\rho \approx 1.8$ is sufficient if $\xi = 5$, for example. In contrast, the simple laminar model described in §14.2 would achieve $\beta = 100$ with $\alpha = 15$, only with the much less realistic value of $\rho \approx 9.1$. With the same β and $\xi, \alpha = 10$ can be supported with $\rho < 1$. Furthermore, for $\beta > 70$, $\xi = 5$ and $\alpha = 5$, $\rho = 0.2$ is sufficient.

Turning to the analysis of the role of the distal connections from A_1^* to $\bar{A}_2$, we observe first that $\tilde{x}^*$ is now represented as $r\beta_\xi$ neurons in A_1^*, in r/ξ columns with $\xi\beta_\xi$ in each column. In this model the axons from each column go in equal numbers to β_ξ distinct columns in other areas (rather than the $\eta\sigma$ as in the original laminar model) of which β_ξ/θ are columns in $\bar{A}_2$. Each such column will then receive ξ axons from the representatives of $\tilde{x}^*$ on the average, since there are β_ξ choices of such columns and $\xi\beta_\xi$ choices of such axons.

Hence on the average r neurons in $\bar{A}_1$ are connected between them to r/ξ columns in A_1^* with $\xi\beta_\xi$ neurons in each, and between them these are connected to $(r/\xi)(\beta_\xi/\theta)$ columns in $\bar{A}_2$, with ξ axons from $\tilde{x}^*$ arriving at each such column. Hence if we define β_1 such that $r\beta_\xi\beta_1$ is the number of nodes in $\bar{A}_2$ synapsed at least α times by axons from $\tilde{x}^*$, then the argument used to derive (14.2) gives

$$r\beta_\xi\beta_1 = \frac{r\beta_\xi}{\xi\theta} \times \frac{\chi}{\rho} \sum_{k=\alpha}^{\infty} e^{-\xi\rho} \frac{(\xi\rho)^k}{k!}$$

or

$$\beta_1 = \frac{1}{\theta} \times \frac{\chi}{(\xi\rho)} \sum_{k=\alpha}^{\infty} e^{-\xi\rho} \frac{(\xi\rho)^k}{k!} = \frac{\beta_\xi}{\theta}. \tag{14.9}$$

We conclude that if α neighbors are required to fire a neuron, then the r representatives of $\tilde{x}$ can make fire $r\beta_\xi^2/\theta$ neurons in $\bar{A}_2$, exactly as in the analysis of (14.3), except now we define β_ξ as we defined β in (14.2) but with ρ replaced by $\xi\rho$. Hence the condition (14.4) is still sufficient to support JOIN, except now much lower values of ρ suffice for the same α, or much higher values of α for the same ρ.

We have, therefore, shown that JOIN can be implemented on the columnar model for less stringent values of α or ρ than needed in the simpler laminar model. The same analysis shows that LINK can be supported similarly. Hence it follows that all vicinal algorithms based on these primitives can also be implemented. As mentioned before, values of α as high as 15 suffice to have $\beta_\xi = 100$ with $\xi = 5$ while maintaining $\rho < 2$. With the same parameters, $\alpha = 10$ can be supported with $\rho < 1$.

14.4   Sparser Random Graphs

In this section we shall show how JOIN and LINK can be implemented on the random graph model defined in §6.3 even when the edge probability p is much smaller than the value $(\mu N/r)^{1/2}/N$ considered there. For example, if $p = (\mu N/r)^{1/4}/N$, then the implied algorithms are still plausibly simple. They can be supported, however, on graphs with an expected number of as few as 168 edges to and from each node, if we use the estimates $N = 10^{10}, r = 50$ and $\mu = 4$. The assumption of $p = (\mu N/r)^{1/2}/N$ of §6.3 would imply a degree of about 2.8×10^4.

First we shall summarize the properties of random graphs that we need. Let $G = (V, E)$ be a directed graph drawn according to such a probability distribution. We let $N \to \infty$ and consider the edge probability p to be a function of N such that $p(N) \to 0$ as $N \to \infty$. We say that node i is at *distance* k from node j if there is a path from j to i containing k edges, but no path containing fewer than k edges. Then we define $\Gamma_{k,r}$ to be the expected number of nodes that are at distance k from at least one member of a randomly chosen set r nodes in V.

Bounds on $\Gamma_{k,r}$ can be proved by adapting some closely related results from the theory of random graphs (B. Bollobás 1981). Note that the expected out-degree of (i.e. edges directed away from) any node is pN. From any one node, naively, one would expect there to be about $(pN)^2$ nodes at distance 2, $(pN)^3$ at distance 3, or, in general $(pN)^k$ nodes at distance k as long as $(pN)^k$ is rather less than N. It can be shown, that under suitable conditions, this is indeed the case, and can be generalized to r nodes rather than one. The result that can be proved is that for any constants r and κ, if $(pN)^{\kappa-1} = \mathrm{o}(N)$ then

$$\Gamma_{k,r} = r(pN)^k(1 + \mathrm{o}(1)) \qquad (14.10)$$

as long as $1 \le k \le \kappa$ and $pN/\log N \to \infty$.

We shall use this result for the case that r is the replication factor, κ is an even constant (e.g. $\kappa = 4$), and $p = (\mu N/r)^{1/\kappa}/N$ for some constant μ.

First we show how to implement JOIN. For this we use a technique we call *train attenuation*. For signaling among neuroids we shall use sequences of firings with short intervals between the firings (e.g. the intervals have the same order of length as each firing). In real neurons this corresponds to trains of spikes traveling down an axon. In order to realize JOIN we will have each node in $\tilde{x}$ and $\tilde{y}$ emit trains of $\kappa+1$ spikes. We shall assume for simplicity that all the nodes in the network can act as relay nodes that behave as follows: whenever a train of j spikes arrives at a node, the node fires $j - 2$ times so as to produce trains of length $j - 2$ along its outputs. The reader can verify that such a program can be easily written for a neuroid that is in a certain waiting state. The result is that only nodes within distance $\kappa/2$ of some member of $\tilde{x}$ or $\tilde{y}$ will be influenced at all. Moreover a node can detect whether it is exactly at distance $\kappa/2$ from a relevant node by detecting that it has received a train of length one.

Our implementation of JOIN will allocate to $z = x \wedge y$ those neuroids $\tilde{z}$ that are at distance exactly $\kappa/2$ from some node in $\tilde{x}$ and from some node in $\tilde{y}$, and are not within a shorter distance of any of these nodes. Exactly as in Algorithm 7.2, which is a vicinal algorithm for this same problem, we will break symmetry between $\tilde{x}$ and $\tilde{y}$ by having them fire at sufficiently separated times that the trains coming from $\tilde{x}$ and $\tilde{y}$ cannot be confused. We note that we can make sure that the presence of more than one path to a $\tilde{z}$ node from the $\tilde{x}$ nodes (say) can be accommodated by the algorithm.

For $\tilde{x}$ and $\tilde{y}$ we define the node sets $\tilde{x}^{(\kappa/2)}$ and $\tilde{y}^{(\kappa/2)}$ to be the nodes at distance $\kappa/2$ from some nodes of $\tilde{x}$ and $\tilde{y}$, respectively. Using the pristine conditions assumption and relation (14.10) one can deduce that the expected size of $\tilde{x}^{(\kappa/2)} \cap \tilde{y}^{(\kappa/2)}$ is $\mu r(1+o(1))$. In the implementation of LINK described below we need that μ be a constant slightly larger than one (e.g. $\mu = 4$). Hence in this implementation of JOIN we will have each node in $\tilde{x}^{(\kappa/2)} \cap \tilde{y}^{(\kappa/2)}$ first recognize itself as such, by going into a special state. From there it will make a random decision that with probability μ^{-1} will take it to a state that will make it behave as a member of $\tilde{z}$, and with probability $1 - \mu^{-1}$ will make it revert to pristine state.

Now we turn to the problem of implementing LINK$(\tilde{x}, \tilde{z})$. We shall use train attenuation here again, but this time to have the nodes in $\tilde{x}^{(\kappa-1)}$ recognize themselves. For example, we can have the nodes in $\tilde{x}$ emit trains of length $2\kappa - 1$, with intermediate nodes always retransmitting trains of length shorter by two than the ones they detect. Once the nodes in $\tilde{x}^{(\kappa-1)}$ have detected themselves, a second stage starts where, as a result of an appropriate schedule of firings of $\tilde{z}$ and $\tilde{x}^{(\kappa-1)}$, any edges from $x^{(\kappa-1)}$ to $\tilde{z}$ will be given some high weight, say 10.

We need to show that the probability that any fixed node in $\tilde{z}$ is not adjacent to any node in $\tilde{x}^{(\kappa-1)}$ is small. Using the pristine conditions and some independence assumptions it is easy to show that any fixed node in $\tilde{z}$ has an edge from an expected number of $\mu + o(1)$ members of $\tilde{x}^{(\kappa-1)}$, and the probability that it has no such edge is $e^{-\mu}(1 + o(1))$, if μ is a constant.

Hence, if $\mu = 4$, for example, and the $o(1)$ term is ignored, we get that whenever $\tilde{x}$ fires so will an expected fraction of $1 - e^{-4} \approx 0.98$ of the nodes of $\tilde{z}$.

In order to establish the correctness of LINK we need to verify an additional condition, namely that the changes made in the weights by the algorithm will not inadvertently cause $\tilde{z}$ to fire when $\tilde{x}$ is not firing but some other items are. Such a calculation can be made in the same manner as for the laminar model in §14.2.

14.5 Another Columnar Model

Lastly, we shall consider a simple model that is inspired by the column structure of the cortex and is very similar to one proposed previously by Braitenberg (V. Braitenberg 1978). Its advantage is that it allows for implementations that are almost as simple as those on the random graph model considered in earlier chapters, but can additionally realize algorithms such as Algorithm 7.2 that exploit bidirectionality, without any assumptions being necessary about the bidirectionality of the long distance connections. Its disadvantages are the same as those of the basic random graph model, that the degree grows as $N^{1/2}$ and that, when interpreted in terms of the cortex, the axons from one small contiguous area go uniformly to all the other areas.

The model is defined probabilistically for N node directed graphs as follows: For an appropriate number m, take m disjoint complete directed subgraphs of N/m nodes each (i.e. in each subgraph each pair of nodes has connections in each of the two directions). Then each of the N nodes will lie in one of these subgraphs. For each such node i, choose one of the complete subgraphs at random (other than its own, say) and add directed edges from i to each one of the N/m nodes in the chosen subgraph. To define this graph class it remains to specify m. We shall choose $m = (Nr^{1/2}/\mu^{1/2})^{1/2}$ where r is the replication factor and μ is a small constant.

The intended correspondence with cortical structure is the following. A complete subgraph corresponds to an idealized column in cortex, a column being a set of neurons in close physical proximity that is richly interconnected. The edges from a node to the members of a distinct subgraph correspond to the long distance axon going to a distant cortical region and synapsing, via axonal branchings, to a large number of neurons in the column to which the axon goes. The assumptions that one neuron is connected to many, both within its column as well as in a single distant column, are reasonable since this rich connectivity is only required in small physical regions. The most problematic aspect of the model is that it requires that the axons emanating from a single column go to a large number, $N/m = (\mu^{1/2}N/r^{1/2})^{1/2}$, of distinct and widely distributed columns. Under the numerical assumptions above of $N = 10^{10}$, $r = 50$, and $\mu = 4$, this number would be about 5.3 $\times$ 10^4. Our aim in discussing this model is to suggest a further possible role for column-like structure in the realization of vicinal algorithms, rather than to suggest that the model is realistic.

In order to implement JOIN consider arbitrary sets $\tilde{x}$ and $\tilde{y}$ of r nodes each. Their elements will be assumed distributed with at most one member in each subgraph since, if they are scattered at random among the N nodes, they will be so distributed with overwhelming probability as $N \to \infty$. Let $k_x \in \tilde{x}$ and $k_y \in \tilde{y}$ be arbitrary points in these two sets. We first compute the expected number of pairs of nodes (i, j) such that i, j are reciprocally connected, i is in the same subgraph as k_x and j is in the same subgraph as k_y. This expected number is $(N/m)^2m^{-2}$ since there are N/m choices of each i or j, and m^{-1} is the probability that any one node is connected to any one subgraph. Substituting $m = (Nr^{1/2}/\mu^{1/2})^{1/2}$ gives μ/r as the sought after expected number.

For implementing JOIN$(\tilde{x}, \tilde{y})$ we shall represent $x \wedge y$ by the set of nodes $\tilde{z}$ that consists of the union of all pairs (i, j) defined in the previous paragraph. Assuming that these sets are disjoint for the r^2 choices of (i, j), which they will be with high probability, we can approximate the expected size of $\tilde{z}$ by $2(\mu/r)r^2$ since there are r^2 choices of the pair (k_x, k_y) and for each such pair (μ/r) pairs of nodes are contributed on the average. Hence $\mu = 0.5$ makes this value equal to r and ensures that $\tilde{z}$ is of the correct expected size.

The algorithm to implement JOIN will fire $\tilde{x}$ and $\tilde{y}$ according to some schedule that will make the node pairs (i, j) just described recognize themselves as being reciprocally connected. The weights will then be changed so that i and j act as virtually one node, and the algorithms in Chapter 7 can then be implemented with these nodes as target nodes.

In order to implement LINK$(\tilde{x}, \tilde{z})$ we will again use node pairs $\{i, j\}$ as defined for JOIN. We will need to set up a circuit such that if the r nodes of $\tilde{x}$ fire then so will a large fraction of $\tilde{z}$. Let k_z be any node in $\tilde{z}$. Then the probability that its subgraph is reciprocally connected to some member of $\tilde{x}$ is $1 - (1 - \hat{p})^r$, where $\hat{p}$ is the probability that any fixed pair of subgraphs is reciprocally connected. To obtain a value for $\hat{p}$ we first note that the probability that at least one node in one fixed subgraph is connected to the nodes of a fixed other subgraph is

$$1 - \left(1 - \frac{1}{m}\right)^{N/m} = 1 - \left(\left(1 - \frac{1}{m}\right)^m\right)^{N/m^2} \rightarrow 1 - e^{-(\mu/r)^{1/2}}$$

for the assumed value of m. Hence $\hat{p}$ is the square of this quantity. Hence if we take $r = 50$ and $\mu = 2$, we can approximate $\hat{p}$ by $(1 - e^{-1/5})^2 \approx 0.033$ and $1 - (1 - \hat{p})^r$ by 0.81.

There is, therefore, an algorithm for implementing LINK$(\tilde{x}, \tilde{z})$ that consists of a first stage where the node pairs (i, j) recognize themselves and go into special states, followed by a second phase where weight changes are made as in the vicinal algorithms of Chapter 8. The effect of future firings of $\tilde{x}$ will be to fire an expected fraction 0.81 of $\tilde{z}$ in the case that $\mu = 2$ and $r = 50$. A network based on a value of μ that suffices to implement LINK can realize JOIN also. As noted in previous chapters, this can be done by invoking JOIN and then randomly freeing an appropriate fraction of the nodes allocated by JOIN so as to simulate the lower value of μ appropriate to this latter function.

Chapter 15
Afterword

Speculation about the nature of the mind is an enduring human preoccupation. For this reason it may be presumptuous for any one generation to believe that it can add significantly to all that has been said by its predecessors. Nevertheless, intellectual opportunities may arise from time to time that offer hope of progress even on this question. In this volume we have advocated the view that the study of appropriately flexible but detailed computational models comprises one such opportunity. For the first time, we now collectively have substantial experience with large scale computations, and, more significantly, have had a few decades to reflect on their possibilities and limitations. Consequently we may have an unprecedented opportunity to make progress. The magnitude of the opportunity is enhanced, of course, by advances in cognitive psychology and neurobiology. It is the results of experiments in these fields that will provide the empirical data which any ultimate theory will have to fit. The theories needed, however, are fundamentally computational and we will first have to gain some understanding of their expected general nature.

The author's interest in this field was sparked more than a decade ago by simple curiosity. Were plausible theories of cognition based on plausible models of neural computation now within our grasp? The major stumbling block at first seemed to be the philosophical problem raised by inductive learning, an aspect of cognition that seems impossible to evade. We believe now that computational learning theory gives an adequate view on this. It explains how it may be possible for a system to learn to cope in a world that is too complex for it to describe or understand. The concrete task that then remained to be carried out was that of assembling and explaining a range of specific functionalities that could be argued both to be central to cognition and to address a significant fraction of its phenomena. While a good theory need not be a theory of everything, in this field it needs to be a theory of enough things. For this reason the range of functionalities that we chose and on which this volume is based is quite broad. It embraces memorization and learning in its various forms. It allows knowledge to be acquired incrementally and to be hierarchical. It includes aspects of knowledge representation that are variously propositional, multi-object, and relational. It takes a view on reasoning. A computational theory of the mind has to address all these issues, and more.

What has been described in this book is a computationally specific theory of such a set of cognitive tasks. It is offered in the spirit of being a first theory that follows our methodology. If it contributes to progress it will do so by encouraging the development of a range of theories of equal or preferably even greater specificity, and experimentation to resolve among them. One of the levels at which it will be appropriate ultimately to evaluate our approach, therefore, will be at this most general methodological level.

How probable is it that beyond the general methodology, some of the specifics described here are correct? The cognitive functions selected were chosen both for being demonstrably implementable on the neuroidal model, as well as for capturing what we believe to be central aspects of cognition. Since computational capabilities are often robust over large ranges of related models, it is a distinct possibility that our set of functionalities is close to some of those implemented by the brain, even if the details of their implementation as described here are not. The accuracy with which the functions that are explained model real cognition offers a second and more demanding level at which a theory of the mind can be evaluated.

Evaluation can and should be done also at the even more detailed levels of the knowledge representation and the algorithms, which may be regarded as the third and fourth levels in this scheme. Since there are usually many algorithmic solutions to any problem that has at least one, it is quite probable that even if the functions and knowledge representation used in a theory are accurate, the actual algorithms are not. We suspect that the range of viable knowledge representations is much more limited than the range of viable algorithms, and, therefore, it will be easier to make progress on understanding the former. In this connection, we note that the representation we have been using here is specific in nature, and fairly easy to characterize. We represent each item of knowledge by a set of neurons, there being no requirement that these neurons be interconnected in any special way. The overall representation is hierarchical and structured. The items represented are sometimes very specific, for example, referring to just one event, and sometimes very general, applicable to a wide variety of objects or events. Following the presentation of an input neurons representing several such items may be triggered. The items will be those that correspond semantically to various aspects of the input and at various levels of generality. Not all combinations of basic concepts are assigned neurons, only those that have been experienced and noticed by the attentional mechanisms. The relationships among the concepts represented can be approximated by Boolean functions, but their boundaries are made less sharp by several factors, such as the fact that the connections among the neurons that are needed ideally, are present only with a certain probability.

The parts of our theory that are specific do not attempt to provide a theory of everything, and even depend on other parts that we describe only informally. Explanations to a similar level of specificity are needed for all the functions we discussed such as attention, resolving among competing concepts, low level vision, and imagery peripherals, as well as possibly others that we have omitted to consider.

In all these cases the resulting theories may be judged at the four levels described. The lowest, the algorithmic level of explanation, is always the most irksome. While it is needed indispensably to support the theory, it is the part that is likely to involve the most unconstrained guesswork. Each algorithm in this text should be viewed primarily, therefore, as a claimed proof that the corresponding task is within the computational capabilities of the brain, rather than as a conjecture about the actual mechanisms used by the brain.

With any theory of the mind there arises the question of whether any progress toward improved machine simulations of human capabilities is implied. The picture of cognition that emerges from our algorithms, while overlapping with that of other authors in various significant respects, is nevertheless distinctive. It may be sketched as follows: Acquired knowledge is represented as a large number of circuits or reflexes that may interact with each other. Each one may be able to take as input both the output from other circuits, as well as information from the sensory or other peripherals. These reflexes are acquired either by memorization or by inductive learning, using data that is either input from the outside, or deduced internally. There is no general mechanism for checking that they are globally consistent with each other. If harmful inconsistencies are detected new reflexes are learned, or internally derived, to resolve between them. The response to a new input is expressible in terms of these previously acquired reflexes. While there is always a capability for learning new notions, once some have been acquired there may be a tendency for the older ones to ramify. From the same starting point such a system may evolve along widely different paths as a result of divergent experiences. Some preprogrammed expertise may also play a crucial role, in the form of algorithms for implementing attention, reflex reasoning, conflict resolution, as well as possibly other generic tasks. Preprogrammed knowledge more directly related to the particular environment of the system, for example, for interpreting the three dimensional world, may also be necessary. The algorithms we have described in this volume can be used to give substance to a small part of this overall picture but clearly much remains to be done before a plausible completion of it can be attempted. We suspect, nevertheless, that constructing artificial intelligence systems that incorporate the range of functionalities that we have shown to be supported by the neuroidal model, would already be of some interest. We consider it particularly encouraging that this model suggests a plausible integrated view of learning and reasoning.

It is perhaps appropriate to conclude by recalling that the efforts made in recent centuries toward understanding the relationship between mind and brain were preceded by a lengthly debate about whether such a relationship existed at all. In the ancient world mental functions were often associated with the heart rather than the brain. Aristotle, whose father had been a physician and who himself contributed much to biology, supported this view. His teacher Plato had believed for theoretical reasons that the head was the true location of the mind. The sphere was the most perfect shape and the head resembled a sphere. An Egyptian papyrus, from more than a millennium earlier had detailed the effects of injuries to various parts of the head on the functions of other, sometimes distant, parts of the body. It was not until a few centuries ago that some consensus on the modern view of the functions of the brain had emerged. It should be encouraging to present day scientists that in the midst of this confusion, and at an early date, some understanding of this question had been reached and expressed with eloquence. Approximately contemporaneously with Plato, Hippocrates — or perhaps one of his followers — wrote[33]: ". . . from the brain, and from the brain only, arise our pleasures, joys, laughter and jests, as well as our sorrows, pains, griefs and tears. Through it, in particular, we think, see, hear, and distinguish the ugly from the beautiful, the bad from the good, the pleasant from the unpleasant, in some cases using custom as a test, in others perceiving them from their utility . . . In these ways I hold that the brain is the most powerful organ of the human body, . . . the brain is the interpreter of consciousness."

becomes tractable after some preliminary training in the course of which these regions are differentiated by color (R.J. Herrnstein *et al.* 1989).

[16]The possible benefits of allowing randomizing steps as a generic resource in algorithms was first emphasized in (M.O. Rabin 1976). In this volume we use randomization for only this one limited purpose, but more general applications in neuroidal algorithms can be easily imagined.

[17]In order to specify an arbitrary function of a set of arguments each taking one of a discrete set of values, it is enough to list all combinations of values that the arguments can take, and give for each one the corresponding value of the function. In the case of $\lambda(s_i, w_i, w_{ji}, f_j)$, f_j can have two values, s_i some constant number, while w_i and w_{ji} can take at most N to some fixed constant power. Hence the product of these four bounds is bounded by a constant power of N as N grows.

[18]The yellow Volkswagen example is from (C.S. Harris 1980).

[19]In all asymptotics where we use the "O" notation that was discussed in §3.1, we shall assume that all quantities are constant except for N, which asymptotes to infinity, and any variables that are defined in terms of N. In this case, since $p = (Nr)^{-1/2}$ and r is a constant independent of N, p asymptotes to 0.

The instance of the Binomial Theorem needed is that for any integer $r \geq 1$ and any number $x, (1 + x)^r = 1 + rx + (r(r - 1)/2)x^2 + \cdots + \binom{r}{i}x^i + \cdots + x^r$. Putting $x = -p$ gives $(1 - p)^r = 1 - rp + O(p^2)$.

For an introduction to probability theory see: W. Feller, *An Introduction to Probability Theory and its Applications*, vol 1, Wiley.

[20]For $x > 1$, $(1 - \frac{1}{x})^x = e^{-1}(1 + O(\frac{1}{x}))$. To derive the asymptotic value of $(1 - \mu/N + y/N)^{N-2r}$ where $y = O(N^{-1/2})$ we let $1/x = (\mu - y)/N$ so that the expression becomes $(1 - \frac{1}{x})^{x(N-2r)/x} = [e^{-1}(1+O(\frac{1}{x}))]^{(\mu-y)(N-2r)/N} = [e^{-1}(1+O(\frac{1}{N}))]^{\mu+O(N^{-1/2})} = e^{-\mu}(1+ O(N^{-1/2}))$.

[21]See J.L. Carter and M.N. Wegman. Universal classes of hash functions. J. Comput. Syst. Sci. *18* (1979) 143-154.

[22]See for example, (D. Angluin 1992, M. Anthony, *et. al* 1992, M. Kearns and U. Vazirani 1994, L.G. Valiant 1984).

[23]See (L.G. Valiant 1985).

[24]See references in note (22) above.

[25]See (L.G. Valiant 1985, M. Kearns and U. Vazirani 1994). One way of achieving an adaptation of this nature is to replace the weight update "if $f_k = 0$ then $w_{ki} := 0$" by "if $f_k = 0$ then $w_{ki} := \max\{0, w_{ki} - v\}$ else $w_{ki} = \min\{1, w_{ki} + u\}$", where $0 < u < v < 1$ are suitable real numbers. The intention is that if x_k is true in more than a fraction

$v/(u+v)$ of the positive examples then it should be taken as belonging to the true rule.

[26]We shall describe a variant of the perceptron algorithm for learning a threshold function $\sum u_j x_j = \underline{u}\,\underline{x} \geq \theta$ that maintains a hypothesis with a fixed threshold that is nonzero, and has fixed-signs. We assume that (i) the expression for the threshold is normalized so that $\sum u_j^2 = 1$, (ii) for some m, for every example $\underline{e} = (e_1, \cdots, e_n) \in \{0,1\}^n$ presented $|\underline{e}|^2 = \sum e_j^2 \leq m$, and (iii) for some $\delta > 0$ for every example $\underline{e}$ either $\underline{u}\,\underline{e} \geq \theta + \delta$ or $\underline{u}\,\underline{e} \leq \theta - \delta$. We also assume that a function sgn: $\{1, \cdots, n\} \to \{+, -\}$ is given, that assigns to each j a sign that is positive if $u_j > 0$, negative if $u_j < 0$, and either one if $u_j = 0$.

The algorithm will maintain a hypothesis of the form $\sum v_j x_j \geq \theta m/\delta$ where $v_j \geq 0$ if $\text{sgn}(j) = +$ and $v_j \leq 0$ if $\text{sgn}(j) = -$. Initially $v_j = 0$ for every j. Following each example $\underline{e}$ on which the true function and the hypothesis disagree, the v_j will be updated as follows: For j with $\text{sgn}(j) = +$, whenever $\underline{u}\,\underline{e} \geq \theta$ and $\underline{v}\,\underline{e} < \theta m/\delta$ then $v_j := v_j + e_j$, and whenever $\underline{u}\,\underline{e} < \theta$ and $\underline{v}\,\underline{e} \geq \theta m/\delta$ then $v_j := \max\{v_j - e_j, 0\}$. Conversely for j with $\text{sgn}(j) = -$, whenever $\underline{u}\,\underline{e} \geq \theta$ and $\underline{v}\,\underline{e} < \theta m/\delta$ then $v_j := \min\{v_j + e_j, 0\}$, and whenever $\underline{u}\,\underline{e} < \theta$ and $\underline{v}\,\underline{e} \geq \theta m/\delta$ then $v_j := v_j - e_j$.

We shall show that for any sequence of examples, the number on which a misclassification is made and hence an update is required is at most m/δ^2. To prove this we shall show that the quantity $H = |\underline{v} - (m/\delta)\underline{u}|^2$ which initially equals m^2/δ^2 and can never be negative, decreases by at least m as a result of each misclassification.

Let $\underline{v}' = \underline{v} + \underline{x}$ be the new hypothesis vector after an update to the old vector v. Then $\underline{x}$ will have all nonnegative or all nonpositive components, depending on which of the two updates is being performed. But e_j and $|x_j|$ will differ only for those j for which $v_j = 0$, and hence $|\underline{v}\,\underline{x}| = |\underline{v}\,\underline{e}|$. (Here we are using the fact that v_j has an integral initial value and is updated by $+1$ or -1, so that it must go through 0 when it changes sign). If we define c to be the increase that H undergoes in an update, then

$$
\begin{aligned}
c &= (\underline{v} - (m/\delta)\underline{u} + \underline{x})^2 - (\underline{v} - (m/\delta)\underline{u})^2 \\
&\leq 2\underline{v}\,\underline{x} - 2(m/\delta)\underline{u}\,\underline{x} + m.
\end{aligned}
$$

Case(i): If $\underline{v}\,\underline{e} < \theta m/\delta$ and $\underline{u}\,\underline{e} \geq \theta + \delta$ then $\underline{x} \geq 0$. Hence $\underline{v}\,\underline{x} = \underline{v}\,\underline{e} < \theta m/\delta$. Also x_j differs from e_j only for j such that $u_j \leq 0$. Let u^* be the sum of all such u_j so that $\underline{u}\,\underline{x} = (\underline{u}\,\underline{e} - u^*)$ and $u^* \leq 0$.

Then

$$c \leq 2\underline{v}\ \underline{e} - (2m/\delta)(\underline{u}\ \underline{e} - u^*) + m$$
$$\leq 2\theta m/\delta - (2m/\delta)(\theta + \delta - u^*) + m$$
$$\leq -m.$$

Case (ii): If $\underline{v}\ \underline{e} \geq \theta m/\delta$ and $\underline{u}\ \underline{e} \leq \theta - \delta$ then $\underline{x} \leq 0$. Hence $\underline{v}\ \underline{x} = -\underline{v}\ \underline{e} \leq -\theta m/\delta$. Also x_j differs from $-e_j$ only for j such that $u_j \geq 0$. Let u^* be the sum of all such u_j so that $\underline{u}\ \underline{x} = -(\underline{u}\ \underline{e} - u^*)$ and $u^* \geq 0$. Then

$$c \leq -2\underline{v}\ \underline{e} - (2m/\delta)(-(\underline{u}\ \underline{e} - u^*)) + m$$
$$\leq -2\theta m/\delta + (2m/\delta)(\theta - \delta - u^*) + m$$
$$\leq -m.$$

[27]Littlestone's upper bound on the number of mistakes for $\alpha = 1 + \delta/2$ is $8n/(\delta^2\theta) + (5/\delta + (14/\delta^2)\ln\theta)\sum_{j=1}^{n} u_j$. The bound grows inversely as δ^2 and becomes very large only if δ becomes vanishingly small (i.e. if the threshold separates the positive and negative examples by very little). If the function being learned is a disjunction then $\delta = 1$.

[28]See (R. Paturi, S. Rajasekaran and J.H. Reif 1989).

[29]Variable binding in recognition, when formulated with no restrictions, is NP-complete and therefore according to general current conjecture, requires exponential time, even if just one binary relation **r** is allowed. To see this consider an undirected graph on n nodes, with the nodes designated as objects $a_1, \cdots, a_n$, and the relation $\mathbf{r}(a_i, a_j)$ defined to be true if and only if there is an edge from node a_i to node a_j in the graph. Then the truth of the expression

$$\exists i_1 < i_2 < \cdots < i_k \bigwedge_{1 \leq j < l \leq k} \mathbf{r}(a_{i_j}, a_{i_l})$$

is equivalent to testing whether the graph has a k-clique, which is known to be NP-complete. A k-clique is a set of k nodes each pair of which is connected by an edge.

[30]See R. Khardon and D. Roth, Learning to reason, manuscript, Harvard University, 1994.

[31]L. Shastri's use of the expression *reflexive reasoning* (L. Shastri and V. Ajjanagadde 1993) also refers to reasoning that is fast. Our use

of the word reflex differs in that we emphasize both that reflexes are typically simple modules, and also that mechanisms need to be exhibited for learning them.

[32] We are using $\binom{u}{i} = \frac{u!}{i!(u-i)!} \leq u^i/i!$.

[33] Hippocrates. The Loeb Classic Library, Vol 2: The *Sacred Disease*, pp 175-179, Harvard University Press, 1959.

Exercises

The exercises below are intended to amplify some of the issues mentioned in the text but not explored there in detail. The first digit in the numbering references the chapter to which the exercise relates most directly.

5.1   Define a neuroid and some initial conditions for it at time $t = 0$, such that if $t = t'$ is the first time some of its inputs fire, then it will fire for the first time at time $t' + 2$.

5.2   Define a neuroid and some initial conditions for it at time $t = 0$, such that if $t = t_o, t_1, t_2, \cdots$ are the times that some of its inputs fire, then $t_0 + 2, t_1 + 2, t_2 + 2, \cdots$ are the times when it will fire.

5.3   Define a neuroid and some initial conditions for it at time $t = 0$, such that it will fire at time $t + 2$ for each t such that some input fires at time $t - 2$ but none fires at time $t - 1$.

5.4   Given a neuroidal net where the states may have various latencies, how could the neuroids be *reprogrammed* so that they all have latency one? By reprogramming we mean the programming of another neuroidal net that computes the same function, but possibly with a different coding of the inputs and the outputs.

5.5   Suppose that we have a neuroidal net NN and wish to construct another one NN^* that simulates it at half speed, and represents each single firing as a succession of two firings. First construct a neuroid that behaves as follows: if it receives inputs at successive times t and $t + 1$, it will fire at times $t + 2$ and $t + 3$, and otherwise it does not fire. How would you perform the general conversion of NN to NN^*.

5.6   Given a neuroidal net in the standard model where λ may depend on w_i, how could it be reprogrammed to remove this dependency?

5.7   In the standard model w_{ji}, q_i and T_i can all be updated at instants when i is not firing. How could a neuroidal net in this model be reprogrammed so that w_{ji} and T_i can be updated only when i is firing?

5.8   Suppose that we allow W to be any finite set, and λ to be any function of its arguments. Show that any neuroidal net expressed in this model can be reprogrammed so that λ does not depend on the mode s_i.

7.1   Suppose that the nodes of any one item $\tilde{x}$ can be caused to fire together, with no other nodes firing. Describe an algorithm that reduces to zero the weights of any edges that go from an $\tilde{x}$ node to any node to which another edge goes from $\tilde{x}$. In what way does this make the allocation of further items easier?

7.2   Write out explicitly the λ and δ transitions for (i) Algorithm 7.1 and (ii) Algorithm 7.2.

7.3   Adapt Algorithm 7.1 so that it works for neuroids in which the allowed values of the threshold cannot be changed.

7.4   Adapt Algorithm 7.2 so that it works for neuroids in which the allowed values of the threshold is restricted to the interval [1,2] of real numbers.

8.1   Suppose that we allocate $\tilde{x}$ as a result of a series of hierarchical allocations by unsupervised memorization. In other words $\tilde{x}$ is at the apex of some such hierarchy. Suppose that we now wish to learn at $\tilde{x}$ a conjunction by supervised memorization. What problems arise if we use the algorithms of Chapters 7 and 8 directly?

8.2   Suppose that $\tilde{x}$ is not allocated hierarchically, but controlled directly by the peripherals. Suppose, however, that at $\tilde{x}$ we wish to memorize in supervised mode a conjunction of items that were previously learned, that may contain disjunctions. What problems arise if we use the algorithms of Chapters 7 and 8 directly?

8.3   Construct a scheme that allows for the memorization of the name for an item x in unsupervised mode, and subsequently the memorization of a conjunction for it in supervised mode, such that at subsequent times either the name or the conjunction will fire appropriate sets of nodes, but no partial mixture (mixed media) of the two kinds of inputs will fire any of them.

8.4   Adapt Algorithm 8.1 so that it achieves a nongraded memorization of a supervised 2-conjunction $z = x_1 \wedge x_2$. In particular, the final state of each $\tilde{z}$ node needs to indicate whether $x_1 \wedge x_2$ has been successfully memorized at that node.

8.5 Mnemonists use a variety of tricks to enable them to memorize apparently arbitrary sequences of almost arbitrary length for long periods. One such trick for memorizing a list of famous people is to imagine each one standing in front of a different house in one's street, the nth person in the list in front of the nth house. Discuss how this method would be implemented on the neuroidal model, and suggest some alternative methods.

9.1 Pavlovian conditioning is a phenomenon in which an animal can be conditioned to produce a response R to an apparently unrelated stimulus A. First a stimulus S and a response R are chosen such that S naturally elicits R (e.g. S can be a puff of air in the eye, and R a blink.) The unrelated stimulus A, the sight of a green circle, say, is presented repeatedly to the animal in conjunction with stimulus S, eliciting R each time as expected. The basic phenomenon of conditioning is that the presentation of A by itself will eventually elicit R even in the absence of S. A second phenomenon, called extinction, occurs if subsequent to this basic training process A is presented a large number of times in the absence of any stimulus that would evoke R. It is found that A will then cease to elicit R by itself. A third phenomenon, called inhibition, also starts with the basic training process, but subsequently randomly mixes presentations of A and S with presentations of A and B where B is a further stimulus that does not elicit R by itself. It is found that after sufficient training A will continue to elicit R, but A and B together will not. Give a neuroidal explanation that accounts for all three of these phenomena. (See (J.E. Mazur 1990) for further phenomena. Note also that inhibitory effects are believed to be implemented as pathways separate from the excitatory ones, and this may be a reflection of the fact that individual neurons are either excitatory or inhibitory.)

9.2 Consider the elimination algorithm described in Section 9.4. Explain how it can be modified so that predicates eliminated by negative examples are never reintroduced as candidate predicates by a positive example.

9.3 We shall say that an inductive algorithm is *teachable by s ex-*

amples, if there exists a sequence of s examples (carefully chosen by a teacher!) that drive the algorithm to the correct hypothesis. For each of the three algorithms (i) elimination for conjunctions, (ii) elimination for disjunctions, and (iii) winnow2, discuss how many examples are required for training them.

10.1   Consider a neuroidal net that at different nodes can support Algorithm 7.2 for unsupervised memorization, Algorithm 8.1 for supervised memorization, winnow2 of Chapter 9, as well as the correlation detection mechanism of Chapter 10. Describe some set of initial conditions that are sufficient to support these four functions together. (No detailed quantitative analysis of the numbers of the various kinds of nodes in necessary.) Describe the condition of the circuit after some learning in each of the four modes has been accomplished.

11.1   Consider a neuroid i that has two nonzero weights $w_{j,i}$ and $w_{k,i}$, both equal to one. Describe an algorithm for it that will make it perform timed conjunctions in the following sense: It will fire at time $t + 2$ if and only if exactly one of j or k fires at time t, and the other one fires at time $t + 1$.

11.2   Describe a neuroid that has many inputs $i_1, \cdots, i_m$ and can be programmed by means of a suitable input schedule, to become equivalent to the timed conjunction neuroid described in the previous exercise, for any pair of its inputs.

References

[M. Abeles 1982] Local Cortical Circuits: An Electrophysiological Study. Springer-Verlag, Berlin.

[M. Abeles 1991] Corticonics: Neural Circuits of the Cerebral Cortex. Cambridge University Press.

[A.V. Aho, J.E. Hopcroft and J.D. Ullman 1974] Design and Analysis of Computer Algorithms, Addison-Wesley, Reading, MA.

[E.J. Aiton 1985] Leibniz: A Biography. Hilger, Bristol, U.K..

[D. Angluin 1992] Computational learning theory. In Proc. 24th ACM Symp. on Theory of Computing, ACM Press, 351-369.

[J.R. Anderson 1981] Cognitive Skills and their Acquisition. (J.R. Anderson, ed.) Erlbaum, Hillsdale, NJ.

[M. Anthony and N. Biggs 1992] Computational Learning Theory. Cambridge University Press.

[H.B. Barlow 1972] Single units and sensation: a neuron doctrine for perceptual psychology. Perception *1*:371-394.

[H.B. Barlow 1985] Central cortex as a model builder. In Models of the Visual Cortex. D. Rose and V.G. Dobson (eds.), Wiley, New York.

[E.B. Baum 1990] The perceptron algorithm is fast for nonmalicious distributions. Neural Computation *2*:248-260.

[F.C. Bartlett 1932] Remembering: A Study in Experimental Social Psychology. Cambridge University Press.

[E. Basar and T.H. Bullock 1992] Induced Rhythms in the Brain. (E. Basar and T.H. Bullock, eds), Birkhäuser, Boston.

[O. Bernander, R.J. Douglas, K.A.C. Martin and C. Koch 1991] Synaptic background activity influences spatiotemporal integration in single pyramidal cells. Proc. Natl. Acad. Sci. USA *88*:11569-11573.

[B. Bollobás 1981] The diameter of random graphs. Trans. Amer. Math. Soc. *267*(1):41-52.

[G. Boole 1854] The Laws of Thought. George Boole's Collected Logical Works, Vol 2, The Open Court Publishing Co., La Salle, Ill., (1952).

[M. Bowermann 1977] The acquisition of word meaning: an investigation of some current concepts. In Thinking: Readings in Cognitive Science, P.N. Johnson-Laird and P.C. Wason (eds.), Cambridge University Press.

[J. Bowers and D.L. Schacter 1994] Priming of novel information in amnesic patients: Issues and data. In Implicit Memory: New Directions in Cognition, Development and Neuropsychology (P. Graf and M. Masson, eds.) Erlbaum, Hillsdale, NJ.

[V. Braitenberg 1978] Cell assemblies in the cerebral cortex. In Theoretical Approaches to Complex Systems, Lecture Notes in Biomathematics 21 (R. Heim and G. Palm, eds.) 171-188.

[V. Braitenberg and A. Schüz 1991] Anatomy of the Cortex: Statistics and Geometry. Springer-Verlag, Berlin.

[J.H. Breasted 1991] The Edwin Smith surgical papyrus. Chicago University Press (2 vols.) .

[R.A. Brooks 1991] Intelligence without reason. Proc. Int. Joint Conf. on Art. Intelligence, 569-595.

[J.S. Bruner, T.J. Goodnow and G.A. Austin 1956] A Study of Thinking. Wiley, New York.

[G.A. Carpenter and S. Grossberg 1987] A massively parallel architecture for a self-organizing neural pattern recognition machine. Computer Vision, Graphics and Image Processing, 37:54-115.

[P.W. Cheng and K.J. Holyoak 1985] Pragmatic reasoning schemas. Cog. Psych. 17:391-416.

[M.T.H. Chi, R. Glaser, M.J. Farr 1988] The Nature of Expertise. Erlbaum, Hillsade, NJ.

[P.S. Churchland and T.J. Sejnowski 1992] The Computational Brain. MIT Press, Cambridge, MA.

[A. Collins and E.E. Smith 1988] Readings in Cognitive Science (A. Collins and E.E. Smith, eds.), Morgan Kaufmann, San Mateo, CA.

[M.L. Commons, et al. 1990] M.L. Commons, R.J. Herrnstein, S.M. Kosslyn and D.B. Mumford. Quantitative Analyses of Behavior, (M.L. Commons, R.J. Herrnstein, S.M. Kosslyn and D.B. Mumford, eds.) Erlbaum, NJ.

[S.A. Cook 1971] The complexity of theorem proving procedures. Proc. Third ACM Symp. on Theory of Computing, ACM Press, 151-158.

[K.J.W. Craik 1943] The Nature of Explanation. Cambridge University Press.

[F. Crick and C. Koch 1990] Towards a neurobiological theory of consciousness. Seminars in Neuroscience 2:263-275.

[S.J. DeArmond, M.M. Fusco and M.M. Dewey 1976] Structure of the Human Brain, Oxford University Press, New York.

[J. DeFelipe and E.G. Jones 1988] Cajal on the Cerebral Cortex, Oxford University Press, New York.

[R. Desimone, T.D. Albright, C.G. Gross and C. Bruce 1984] J. Neuroscience, 4(8):2051-2062.

[R.O. Duda and P.E. Hart 1973] Pattern Classification and Scene Analysis. Wiley, New York.

[J.C. Eccles 1989] Evolution of the Brain: The Creation of the Self. Routledge, London and New York.

[J.St.B.T. Evans 1983] Thinking and Reasoning. (J.St.B.T. Evans, ed.), Routledge, London.

[M.W. Eysenck 1984] A Handbook of Cognitive Psychology, Lawrence-Erlbaum, Hillsdale, NJ.

[J.A. Feldman 1982] Dynamic connections in neural networks. Biol. Cybern. 46:27-39.

[J.A. Feldman 1990] Computational constraints on higher neural representations. In Computational Neuroscience, (E.L. Schwartz, ed.), MIT Press.

[J.A. Feldman and D.H. Ballard 1982] Connectionist models and their properties. Cog. Sci. 6:205-254.

[D.J. Felleman and D.C. van Essen 1991] Distributed hierarchical processing in primate cerebral cortex. Cerebral Cortex 1:1-48.

[W.J. Freeman 1975] Mass Action in the Nervous System. Academic Press.

[C.R. Gallistel 1990] The Organization of Learning. MIT Press, Cambridge, MA.

[M.R. Garey and D.S. Johnson 1978] Computers and Intractability: A Guide to the Theory of NP-Completeness, Freeman, San Francisco.

[A. Gerbessiotis 1993] Topics in Parallel and Distributed Computation, Ph.D. Thesis, Division of Applied Sciences, Harvard University, Cambridge, MA.

[M.L. Ginsberg 1989] Readings in Nonmonotonic Reasoning. Morgan Kaufmann, Los Altos, CA.

[M.A. Gluck and D.E. Rumelhart 1990] Neuroscience and Connectionist Theory. (M.A. Gluck and D.E. Rumelhart, eds.), Erlbaum, Hillsdale, NJ.

[P.M. Gochin *et al.* 1991] P.M. Gochin, E.K. Miller, C.G. Gross and G.L. Gerstein. Functional interactions among neurons in macaque inferior temporal cortex. Exp. Brain Res., *84*:506-516.

[M. Golea and M. Marchand 1993] On learning perceptrons with binary weights. Neural Computation *5*:767-782.

[N. Goodman 1983] Fact, Fiction and Forecast. Harvard University Press, Cambridge, MA.

[C.G. Gross 1992] Representation of visual stimuli in inferior temporal cortex. Phil Trans R. Soc. Lond. B *335*:3-10.

[C.G. Gross, *et al.* 1993] C.G. Gross, H.R. Rodman, P.M. Gochin and M. W. Colombo. Inferior temporal cortex as a pattern recognition device. In Computational Learning and Cognition, (E.B. Baum, ed.), SIAM, Philadelphia.

[C.S. Harris 1980] Insight or Out of Sight? In Visual Coding and Adaptability (C.S. Harris, ed.) Erlbaum, Hillsdale, NJ, 95-149.

[D. Haussler 1988] Space efficient learning algorithms. Technical Report UCSC-CRL-88-2. Baskin Center for Computer Engineering and Information Sciences, University of California at Santa Cruz, Santa Cruz, CA.

[D.O. Hebb 1949] The Organisation of Behavior. Wiley, New York.

[R.J. Herrnstein 1985] Riddles of natural categorization. Phil. Trans. R. Soc. Lond. B *308*:129-144.

[R.J. Herrnstein, 1990] Levels of stimulus control: A functional approach. Cognition *37*:133-166.

[R.J. Herrnstein *et al.* 1989] R.J. Herrnstein, W. Vaughan, D.B. Mumford and S.M. Kosslyn. Teaching pigeons an abstract relational rule: Insideness. Perception and Psychophysics, *46*(1):56-64.

[G.E. Hinton, J.L. McClelland and D.E. Rumelhart 1986] Distributed Representations. In (McClelland and Rumelhart).

[M.A. Hofman 1989] On the evolution and geometry of the brain in mammals. Progress in Neurobiology *32*:137-158.

[J.E. Hopcroft and J.D. Ullman 1979] Introduction to Automata Theory, Languages and Computation. Addison Wesley.

[J.J. Hopfield 1982] Neural networks and physical systems with emergent computational abilities. Proc. Nat. Acad. Sci. *79*:2554.

[D.H. Hubel and T.N. Wiesel 1977] Functional architecture of macaque monkey visual cortex. Proc. Roy. Soc Lond. B *198*:1-59.

[H.J. Jerison 1991] Fossil brains and the evolution of neocortex. In The Neocortex: Ontogeny and Phylogeny, (B.L. Finlay, G. Innocenti and H. Scheich, eds.), Plenum Press, New York.

[M.K. Johnson and L. Haster 1987] Human learning and memory. Ann. Review of Psychology, 631-668.

[P.N. Johnson-Laird and P.C Wason (eds.) 1977] Thinking: Readings in Cognitive Science, Cambridge University Press.

[P.N. Johnson-Laird 1983] Mental Models: Towards a Cognitive Science of Language, Inference and Consciousness. Cambridge University Press and Harvard University Press.

[P.N. Johnson-Laird 1988] The Computer and the Mind: an Introduction to Cognitive Science. Harvard University Press, Cambridge, MA.

[D. Kahneman and A. Tversky 1973] On the psychology of prediction. Psych. Rev. *80*:237-251.

[E.R. Kandel, J.H. Schwartz and T.M. Jessell 1991] Principles of Neural Science. 3rd Edition. Elsevier-North Holland, New York.

[A. Karatsuba and Y. Ofman 1962] Multiplication of multidigit numbers on automata. Dokl. Akad. Nauk. SSSR *145:2* (1962) 293-294. (Engl. trans in Sov. Phys. Dokl. 7:595-596.)

[R.M. Karp 1972] Reducibility among combinatorial problems. In Complexity of Computer Computations, (R.E. Miller and J.W. Thatcher, eds.) Plenum Press, New York, 85-104.

[M.J. Kearns and L.G. Valiant 1989] Cryptographic limitations on learning Boolean formulae and finite automata. Proc. 21st ACM Symp. on Theory of Computing, Assoc. Comp. Mach., New York, 433-444.

[M. Kearns and U. Vazirani 1994] Introduction to Computational Learning Theory. MIT Press, Cambridge, MA.

[M. Kharitonov 1993] Cryptographic hardness of distribution-specific learning. Proc. 25th ACM Symp. on Theory of Computing, Assoc. Comp. Mach., New York, 372-381.

[F. Klix and H. Hagendorf 1986] Human Memory and Cognitive Capabilities, vols 1-2, (F. Klix and H. Hagendorf, eds.) North Holland, Amsterdam.

[W. Kneale and M. Kneale 1962] The Development of Logic. Oxford: Clarendon Press.

[C. Koch and I. Segev 1989] Methods in Neuronal Modeling: From Synapses to Networks. (C. Koch and I. Segev, eds.) MIT Press, Cambridge, MA.

[S.M. Kosslyn 1980] Image and Mind. Harvard University Press.

[S.M. Kosslyn and O. Koenig 1992] Wet Mind. Free Press, New York.

[S.M. Kosslyn and R.A. Anderson 1992] Frontiers in Cognitive Neuroscience. (S.M. Kosslyn and R.A. Anderson, ed.) MIT Press, Cambridge, MA.

[K.S. Lashley 1950] In search of the engram: Physiological mechanisms in animal behavior. In Symposium of the Society for Experimental Biology. (J.F. Danielli and R. Brown, eds.) Cambridge University Press, Cambridge.

[H. Levesque 1986] Making believers out of computers. Artificial Intelligence *30*:81-108.

[Y. Levy, *et al.* 1988] Y. Levy, I.M. Schlesinger and M.D.S. Braine. Categories and Processes in Language Acquisition. (Y. Levy, I.M. Schlesinger and M.D.S. Braine, eds.), Erlbaum, Hillsdale, NJ.

[M. Li and P. Vitányi 1993] An Introduction to Kolmogorov complexity and its Applications. Springer Verlag, New York.

[N. Littlestone 1988] Learning quickly when irrelevant attributes abound: a new linear-threshold algorithm. Machine Learning *2*:285-318.

[N. Littlestone 1989a] Mistake Bounds and Logarithmic Linear-threshold Learning Algorithms. Ph.D. Thesis, Univ. of California at Santa Cruz, March 1989.

[N. Littlestone 1989b] From on-line to batch learning. Proc. of Second Annual Workshop on Computational Learning Theory, Morgan Kaufmann, San Mateo, California, 269-284.

[A.R. Luria 1976] The Neuropsychology of Memory. Winston and Sons, Washington, DC.

[A.R. Luria 1980] Higher Cortical Functions in Man. Basic Books, New York.

[P. Maes 1990] Designing Autonomous Agents. (P. Maes, ed.), MIT Press, Cambridge, MA.

[D. Marr 1970] A theory of the cerebral neocortex. Proc. Roy. Soc. Lond. B*176*:161-234.

[D. Marr 1982] Vision. W.H. Freeman, New York.

[A. Mason *et al.* 1991] A. Mason, A. Nicoll, and K. Stratford. J. Neuroscience *11*(1):72-84.

[J.H.R. Maunsell and W.T. Newson 1987] Visual processing in monkey extrastriate cortex. Ann Rev. Neuroscience *10*:363-401.

[J.E. Mazur 1990] Learning and Behavior. Prentice Hall, Englewood Cliffs, NJ.

[J. McCarthy 1980] Circumscription — A form of non-monotonic reasoning. Artificial Intelligence *13*:27-39.

[J.L. McClelland and D.E. Rumelhart 1986] Parallel Distributed Processing. Vols 1-2. MIT Press, Cambridge, MA.

[W.S. McCulloch and W.H. Pitts 1943] A logical calculus of ideas immanent in nervous activity. Bull. of Math. Biophysics, *5*:115.

[M. Minsky 1975] A framework for representing knowledge. In the Psychology of Computer Vision. P.H. Winston (ed.) McGraw-Hill, New York.

[M. Minsky 1986] The Society of Mind. Simon and Schuster, New York.

[M. Minsky and S. Papert 1969] Perceptrons. MIT Press, Cambridge, MA.

[V. Mountcastle 1979] An organization principle of cerebral function. In The Neurosciences: 4th Study Program, (F.O. Schmitt, F.D. Worden, eds.), MIT Press, Cambridge, MA.

[G.A. Murrell and J. Morton 1974] Word recognition and morphemic structure. J. Expt. Psych. *102*(6):963-968.

[K. Nakayama and G.H. Silverman 1986] Serial and parallel processing of visual feature conjunctions, Nature *320*:264-5.

[W.J.H. Nauta and M. Feirtag 1976] Fundamental Neuroanatomy. W.H. Freeman, San Francisco.

[U. Neisser 1982] Memory Observed: Remembering in Natural Contexts. W.H. Freeman, San Francisco.

[A. Newell 1990] Unified Theories of Cognition. Harvard University Press, Cambridge, MA.

[A. Newell and H.A. Simon, 1972] Human Problem Solving. Prentice Hall, Englewood Cliff, NJ.

[J.G. Nicholls, A.R. Martin and B.G. Wallace 1992] From Neuron to Brain, 3rd edition. Sinauer Associates, Sunderland, MA.

[R.E. Passingham 1982] The Human Primate, W.H. Freeman, Oxford.

[R. Paturi, S. Rajasekaran and J.H. Reif 1989] The light bulb problem. Proc. 2nd Ann. Workshop on Computational Learning Theory, Morgan Kaufmann, San Mateo, CA, 261-268.

[I.P. Pavlov 1927] Conditioned Reflexes. Oxford University Press, Oxford.

[I.P. Pavlov 1928] Lectures on Conditioned Reflexes. International Publishers, New York.

[J.M. Pearce 1987] An Introduction to Animal Cognition. Erlbaum, Hove, Sussex, UK.

[R. Penrose 1989] The Emperor's New Mind. Oxford University Press.

[D.H. Perkel *et al.* 1967] D.H. Perkel, G.L. Gerstein and G.P. Moore. Neuronal spike trains and stochastic point processes: II Simultaneous spike trains. Biophys J. 7:419-440.

[A. Peters and E.G. Jones 1984] Cerebral Cortex. Vols 1-8, (A. Peters and E.G. Jones, eds.) Plenum Press, New York.

[S.E. Peterson, *et al.* 1989] S.E. Peterson, P.T. Fox, M.I. Posner, M. Minton and M.E. Raichle, Positron emission tomographic studies of the processing of single words. J. of Cognitive Neuroscience *1*:153-170.

[S.E. Peterson, P.T. Fox, A.Z. Snyder and M.E. Raichle 1990] Activation of extra-striate and frontal cortical areas by visual words and word-like stimuli. Science *249*:1041-1044.

[C.G. Phillips, S. Zeki and H.B. Barlow 1984] Localization of function in the cerebral cortex. Brain *107*:327-361.

[M.I. Posner, S.E. Peterson, P.T. Fox and M.E. Raichle 1988] Localization of cognitive functions in the human brain. Science *264*:1627-1631.

[M.I. Posner 1989] Foundations of Cognitive Science. (M.I. Posner, ed.) MIT Press.

[M.I. Posner and S.E. Peterson 1990] The attention system of the human brain, Ann. Review of Neuroscience, Vol 13: 25-42.

[M.I. Posner and M.K. Rothbart 1991] Attention and Conscious Experience. In The Neuropsychology of Consciousness, (A.D. Milner and M.D. Rugg, eds.), Academic Press, London, 91-112.

[D. Premack 1983] The codes of man and beasts. Behav. and Brain Sci., 6:125-167.

[M.O. Rabin 1976] Probabilistic algorithms. In Algorithms and Complexity: New Directions and Recent Results, (J.F. Traub, ed.) Academic Press, NY, 21-40.

[R. Reiter 1987] Nonmonotonic reasoning. Ann. Reviews of Computer Science, 2:147-187.

[R.A. Rescorla 1980] Pavlovian Second-order Conditioning. Erlbaum, Hillsdale, NJ.

[R.A. Rescorla and A.R. Wagner 1972] A theory of Pavlovian conditioning: Variations in the effectiveness of reinforcement and nonreinforcement. In Classical Conditioning II: Current Research and Theory, (A.H. Black and W.F. Prokasy, eds.), Appleton-Century-Crofts, New York.

[A. Richardson-Klavehn and R.A. Bjork 1988] Measures of memory. Ann. Rev. Psychology 39:475-543.

[L.J. Rips 1990] Reasoning. Ann Rev. Psych. 91:321-353.

[E. Rosch 1977] Classification of real-world objects: origins and representations in cognition. In Thinking: Readings in Cognitive Science. P.N. Johnson-Laird and P.C. Wason (eds.), Cambridge University Press.

[F. Rosenblatt 1958] The perceptron, a probabilistic model of information storage and organization in the brain. Psych. Review, 62:386.

[F. Rosenblatt 1962] Principles of Neurodynamics. Spartan, New York.

[D.E. Rumelhart, G.E. Hinton and R.J. Williams 1986] Learning representations by back-propagating errors. Nature 323:533-536.

[D.L. Schacter 1992] Understanding implicit memory. American Psychologist 47:559-569.

[R.C. Schank and R.P. Abelson 1977] Scripts, Plans, Goals and Understanding. Erlbaum, Hillsdale, NJ.

[D.A. Scholl 1956] The Organization of Cerebral Cortex. Methuen, London.

[E.L. Schwartz 1990] Computational Neuroscience,(E.L. Schwartz, ed.) MIT Press, Cambridge, MA.

[T.J. Sejnowski and C.R. Rosenberg 1987] Parallel networks that learn to pronounce English text. Complex Systems 1:145-168.

[L. Shastri and V. Ajjanagadde 1993] From simple associations to systematic reasoning: A connectionist representation of rules, variables and dynamic bindings using temporal synchrony. Behavioral and Brain Sciences 16(3):417-494.

[G.M. Shepherd 1990] Synaptic Organization of the Brain. 3rd Edition (G.M. Shepherd, ed.), Oxford University Press.

[G.M. Shepherd 1990a] The significance of real neuron architectures for neural networks simulations. In (E. L. Schwartz 90).

[W. Singer 1990] Search for coherence: a basic principle of cortical self-organization. Concepts in Neuroscience *1*:1-26.

[E.E. Smith, C. Langston, and R.E. Nisbett 1992] The case for rules in reasoning. Cog. Sci. *16*:1-40.

[E.E. Smith and D.C. Medin 1981] Categories and Concepts. Harvard University Press.

[K.L. Smola 1932] An objective study of concept formation. Psychological Monographs *42*, No. 191.

[W.R. Softky and C. Koch 1992] The highly irregular firing of cortical cells is inconsistent with temporal integration of random EPSPs. J. Neuroscience, *13*(1):334-350.

[L.R. Squire 1987] Memory and Brain. Oxford University Press, New York.

[L.R. Squire 1992] Encyclopedia of Learning and Memory. (L.R. Squire, ed.) Macmillan, NY .

[L. Standing 1973] Learning 10,000 pictures. Quart. J. Expt. Psych. *25*:207-222.

[M. Sur *et al.* 1988] M. Sur, P.E. Garraghty and A.W. Roe. Experimentally induced visual projections into auditory thalamus and cortex. Science *242*:1437-1441.

[J.G. Sutcliffe 1988] mRNA in the mammalian nervous system. Ann Rev. Neuroscience *11*:157-198.

[R.S. Sutton and A.G. Barto 1981] Toward a modern theory of adaptive networks: expectation and prediction. Psychol. Rev. *88*:135-170.

[J. Szentágothai 1978] The neuron network of the cerebral cortex: a functional interpretation. Proc. R. Soc. Lond. B *201*:219-248.

[K. Tanaka, H.A. Saito, Y. Fukada and M. Moriya 1991] Coding visual images of objects in the inferotemporal cortex of the macaque. J. Neurophysiol. *66*:170-189.

[A.M. Thomson *et al.* 1988] A.M. Thomson, D. Girdlestone, D.C. West. J. Neurophysiology *60*(6):1896-1907.

[A.M. Thomson *et al.* 1993] A.M. Thomson and D.C. West. Neuroscience, *54*(2):329-346. Also A.M. Thomson, J. Deuchars and D.C. West. Neuroscience *54*(2):347-360.

[S.J. Thorpe *et al.* 1989] S.J. Thorpe, K. O'Regan and A. Pouget. Humans fail on XOR pattern classification problems. In Neural Networks from Models to Applications (L. Personnaz and G. Dreyfus, eds.) IDSET, Paris.

[P. V. Tobias 1987] The brain of homo habilis: a new level of organization of cerebral evolution. J. Hum. Evol. *16*:741-761.

[A. Treisman 1988] Feature and Objects: The Fourteenth Bartlett Memorial Lecture. The Quart. J. of Expt. Psych. 40A(2):201-237.

[A. Treisman and G. Gelade 1980] A feature-integration theory of attention. Cognitive Psychology *12*:97-136

[A.M. Turing 1936] On computable numbers, with an application to the Entscheidungsproblem. Proc. Lond. Math. Soc. 2nd Series, *42*:230-265. Corrections, ibid, *43*:544-546.

[A.M. Turing 1950] Computing machinery and intelligence. Reprinted in Minds and Machines (A.R. Anderson, ed.) Prentice-Hall, Englewood Cliffs, NJ 1964.

[E. Tulving 1983] Elements of Episodic Memory. Oxford: Clarendon Press.

[E. Tulving and D.L. Schacter 1990] Priming and human memory systems. Science *247*:301-306.

[L.G. Valiant 1984] A theory of the learnable. Comm. Assoc. Comp. Mach. *27*: 1134-1142.

[L.G. Valiant 1985] Learning disjunctions of conjunctions. International Joint Conf. on Art. Intelligence. Morgan Kaufmann, Los Altos, CA, 560-566.

[L.G. Valiant 1988] Functionality in neural nets. Proc. 7th Nat. Conf. on Art. Intelligence, AAAI, Morgan Kaufmann, San Mateo, CA, 629-634.

[C. von der Malsberg and W. Schneider 1986] A neural cocktail-party processor. Biol. Cybern. *54*:29-40.

[P.C. Wason 1983] Realism and rationality and the selection task. In Thinking and Reasoning: Psychological Approaches, (J. St. B.T. Evans, ed.), Routledge.

[E.L. White 1989] Cortical Circuits. Birkhäuser, Boston.

[J.M. Wolfe, K.R. Cave and S.L. Franzel 1989] Guided search: an alternative to the feature integration model for visual search. J. Expt. Psych: Human perception and performance. *15*(3):419-433.

[C.D. Woody *et al.* 1988] C.D. Woody, D.L. Alkon and J.L. McGaugh. Cellular Mechanisms of Conditioning and Behavioral Plasticity. Plenum Press, New York.

[S. Zola-Morgan and L.R. Squire 1992] Neuroanatomy of memory. Ann. Rev. Neuroscience *16*:547-563.

Index of Notation

A	Available state.
A	A cortical area.
B	Busy state.
C	Conjunctive state.
D	Disjunctive state.
E	Set of edges of a graph.
$E(\tilde{x})$	Set of directed neighbors of node set $\tilde{x}$ in directed graph with edge set E.
Exp( )	Expected value of a random variable.
F	Firing state.
H	A logical expression.
IC	Initial conditions (§5.2).
IS	Input sequence (§5.2).
$\tilde{I}, \tilde{J}$	Sets of neuroids prompted by peripherals.
K	A set.
L	Local state.
M	Memorization state.
M	Number of items stored in the NTR.
M_e	Maximum number of items in the NTR that can be charged simultaneously.
N	Number of neuroids in the NTR or in one area (§5.2 and §14.2).
P	Probabilistic or correlational state.
Q	Set of states.
R	Relay state.
S	Supervised state.
T_i	Value of threshold for neuroid i.
$\underline{T_i}$	Vector of numerical values held by neuroid i.
U	Unsupervised state.
V	Set of nodes of a graph.

231

W	Set of weights allowed on the edges.
X	Set of modes.
a, b, c	Objects in relational expression.
f_i	Firing status of neuroid i.
e	2.71828.....
e_i	A Boolean variable.
$\underline{e}$	Vector $(e_1, \cdots, e_u)$ of Boolean variables.
i, j, k, l, m, n	Nonnegative integer variables, sometimes used to identify individual neuroids from $\{1, 2, \cdots, N\}$.
$\ell(\ )$	Latency of a state.
p	The value of a probability.
q_i	The state of neuroid i.
r	Replication factor in a neuroidal system (§6.1).
s_i	The mode of neuroid i. In general $s_i = (q_i, \underline{T}_i)$.
t	Integer representing time.
u, v	Real numbers.
w_{ij}	Weight on edge from neuroid i to neuroid j.
w_i	Sum of weights of edges from currently firing neuroids to neuroid i.
x, y, z	Items, Boolean variables, predicates, $\{0, 1\}$ real variables.
$\tilde{x}, \tilde{y}, \tilde{z}$	Neuroids representing items x, y, z, respectively.
Γ	Number of nodes within a certain distance of others (§14.4).
α	Minimum number of synapses needed to fire a neuron (§2.3, §14.2).
β	Number of neighbors a neuroid can cause to fire by itself (§14.2).
γ	Number of elements in vector $\underline{T}$.
δ	Update function for mode (§5.2).
δ	Parameter in winnow2.
η	Number of neurons in each cubic millimeter of cortex.
θ	Parameter in linear inequality (§9.5).
θ	Number of areas to which axons from one area project (§14.2).
κ	Ratio N/r (§14.2).
κ	Parameter of graph related to diameter (§14.4).
λ	Weight update function (§5.2).

μ	Constant multiplier in graph density (§6.4, §14.2).
ν	Proportion of representatives of an item that need to fire (§14.3).
ξ	Degree of locality (§14.4).
π	3.14159...
ρ	Measure of connectivity of neurons in close proximity (§14.2).
σ	Volume in cubic millimeters occupied by an axonal branching.
τ_i	Units of time at various scales (§11.3).
ϕ	Number of cortical areas (§14.2).
χ	Number of synapses on an axonal branching (local or distal) (§12.2).
ψ	Number of dendritic trees occupying approximately some volume as one axonal branching (§14.2).
$\|K\|$	Number of elements in set K.
(i,j)	Directed edge from node i to node j.
$\Rightarrow$	Then one time unit later . . . .
log	Logarithm to the base 2.
$\wedge$	Boolean conjunction ($x \wedge y$ is sometimes abbreviated to xy).
$\vee$	Boolean disjunction ($x \vee y$ is sometimes abbreviated to $x + y$).
$\neg$	Boolean negation.
$\equiv$	Logical equivalence of expressions.
$O(g)$	Denotes any function h that grows at most as fast as some constant multiple of the function g, where both h and g are functions of N. Thus for some constant $k \geq 0$, $h \leq kg$ for all $N \geq 0$.
$o(g)$	Denotes any function h that grows more slowly than g, where both h and g are functions of N. Thus for any constant $\varepsilon > 0$ there is an N_0 such that $h \leq \varepsilon g$ for all $N \geq N_0$.
∞	Infinity.
$\{...\}$	Set.
$\in$	Set membership.

Index

All the terms listed here are used in the text in some technical sense. Some are given mathematical definitions or are biological terms. The remainder are specified less precisely, but their inclusion here is intended to emphasize that some specific technical meaning is implied.